Non-Linear Optical Materials

by R. Saravanan

Chapter I provides an introduction to linear optics and the physical origin of non-linear optical phenomena. The principle characterization techniques for analyzing the microstructural, optical and morphological properties of non-linear optical materials are discussed: Powder X-ray diffraction (PXRD), UV-Visible spectroscopy, scanning electron microscopy (SEM), and energy dispersive X-ray spectroscopy (EDS). Also presented are methods for the structural refinement of these materials, as well as the analysis of electron density distribution by means of novel techniques and the corresponding computational procedures.

Chapter II describes sample preparation and PXRD analysis of a number of non-linear optical materials, such as $PbMoO_4$, $LiNbO_3$, $Ce:Gd_3Ga_5O_{12}$, $CaCO_3$, $Yb:CaF_2$, and Al_2O_3, $Cr:Al_2O_3, V:Al_2O_3$.

Chapter III deals with the optical properties and micro-structural characterization of non-linear optical materials, such as $PbMoO_4$, $LiNbO_3$, $Ce:Gd_3Ga_5O_{12}$, $CaCO_3$, $Yb:CaF_2$, and Al_2O_3, $Cr:Al_2O_3, V:Al_2O_3$. The band gap, crystallite size and particle size of these materials are determined by means of UV-visible spectroscopy, powder X-ray profile analysis and scanning electron microscopy. Also discussed is the elemental compositional analysis for $PbMoO_4$, $LiNbO_3$, $Ce:Gd_3Ga_5O_{12}$, $CaCO_3$, $Yb:CaF_2$, and Al_2O_3, $Cr:Al_2O_3, V:Al_2O_3$.

Chapter IV focusses on the electron density distribution analysis of non-linear optical materials, such as $PbMoO_4$, $LiNbO_3$, $Ce:Gd_3Ga_5O_{12}$, $CaCO_3$, $Yb:CaF_2$, and Al_2O_3, $Cr:Al_2O_3, V:Al_2O_3$. The results are presented in the form of electron density maps and profiles. The bonding behavior of these materials is studied using both quantitative and qualitative analysis.

Chapter V centers on the inter-atomic ordering in non-linear optical materials, and presents computations of the pair distribution function (atomic correlation function) for selected materials.

Keywords

Non-linear optical materials, Powder X-ray diffraction, UV-Visible spectroscopy, Scanning electron microscopy, Energy dispersive X-ray spectroscopy, Electron density distribution, Sample preparation, Pair distribution function, Atomic correlation function

Non-Linear Optical Materials

by

Dr. R. Saravanan, M.Sc., M.Phil., Ph.D.

Associate Professor & Head

Research Centre and PG Department of Physics

The Madura College (Autonomous)

Madurai - 625 011

India

Published by **Materials Research Forum LLC**
Millersville, PA 17551, USA

Published as part of the book series
Materials Research Foundations
Volume 28 (2018)
ISSN 2471-8890 (Print)
ISSN 2471-8904 (Online)

Print ISBN 978-1-945291-60-9
ePDF ISBN 978-1-945291-61-6

Distributed worldwide by

Materials Research Forum LLC
105 Springdale Lane
Millersville, PA 17551
USA
http://www.mrforum.com

Manufactured in the United States of America
10 9 8 7 6 5 4 3 2 1

Table of Contents

Preface

Chapter I summarizes the basic introduction to linear optics, and the physical origin of non-linear optical phenomena. It provides and discusses the principles of the characterization techniques such as, powder X-ray diffraction (PXRD), UV-Visible spectroscopy, scanning electron microscopy (SEM), and energy dispersive X-ray spectroscopy (EDS) which have been used for analyzing the structural properties of the synthesized crystal systems, optical properties, morphological properties, and the elemental composition of the prepared non-linear optical materials respectively. This chapter also deals with the methodology of the structural refinement, the analysis of electron density distribution using currently available novel techniques with the corresponding computational procedure in a detailed manner, and the real space analysis of the nearest neighbor distance using atomic correlation functions. A review of the earlier work done on the chosen non-linear optical materials, $PbMoO_4$, $LiNbO_3$, $Ce:Gd_3Ga_5O_{12}$, $CaCO_3$, $Yb:CaF_2$, and Al_2O_3, $Cr:Al_2O_3$, $V:Al_2O_3$ has also been presented in this chapter. It also provides the scope of the present research work.

Chapter II provides the methods of sample preparation and the analysis of the observed powder X-ray diffraction data sets collected using these samples of non-linear optical materials such as, $PbMoO_4$, $LiNbO_3$, $Ce:Gd_3Ga_5O_{12}$, $CaCO_3$, $Yb:CaF_2$, and Al_2O_3, $Cr:Al_2O_3$, $V:Al_2O_3$. The results of fitting PXRD profiles for all the non-linear optical materials are reported and analyzed.

Chapter III deals with the optical properties and micro-structural characterization of non-linear optical materials such as $PbMoO_4$, $LiNbO_3$, $Ce:Gd_3Ga_5O_{12}$, $CaCO_3$, $Yb:CaF_2$, and Al_2O_3, $Cr:Al_2O_3$, $V:Al_2O_3$ in a detailed manner. The band gap, the crystallite size and the particle size of the chosen non-linear optical materials from UV-visible analysis, powder X-ray profile and scanning electron microscope respectively are also determined and discussed in this chapter. This chapter also discusses the elemental compositional analysis for $PbMoO_4$, $LiNbO_3$, $Ce:Gd_3Ga_5O_{12}$, $CaCO_3$, $Yb:CaF_2$, and Al_2O_3, $Cr:Al_2O_3$, $V:Al_2O_3$.

Chapter IV deals with the electron density distribution analysis of non-linear optical materials such as $PbMoO_4$, $LiNbO_3$, $Ce:Gd_3Ga_5O_{12}$, $CaCO_3$, $Yb:CaF_2$, and Al_2O_3, $Cr:Al_2O_3$, $V:Al_2O_3$. The results of electron density distribution studies are presented in the form of three, two dimensional electron density maps and one dimensional profile. This chapter provides quantitative and qualitative analysis of the bonding behaviour of the chosen non-linear optical materials.

Chapter V deals with the study of the inter-atomic ordering i.e., the local structure ordering of the selected non-linear optical materials. The pair distribution function (atomic correlation function), is computed for selected non-linear optical materials of the present work.

A comprehensive analysis of the results of the reported work has been presented in the conclusion section as **Chapter VI**.

Some results of the present analysis have been published as follows;

1. Charge density in MoO_4 tetrahedron and PbO_8 octahedron in $PbMoO_4$, T. K. Thirumalaisamy and R. Saravanan, *Journal of Materials Science: Materials in Electronics,* Vol. 22, pp. 1637-1648, 2011.

2. Local structure determination of the nonlinear optical material $LiNbO_3$ using XRD, R. Saravanan, T. K. Thirumalaisamy and T. Kajitani, *Physica Status Solidi A,* Vol. 208(11), pp. 2643–2650, 2011.

3. The redistribution of charge density in $CaF_2:Yb^{3+}$, T. K. Thirumalaisamy, R. Saravanan and S. Saravanakumar, *Journal of Materials Science: Materials in Electronics,* Vol. 26(9), pp. 6683–6691, 2015.

4. Structure and charge density of Ce doped gadolinium gallium garnet(GGG), T. K. Thirumalaisamy, S. Saravanakumar, Skirmante Butkute, Aivaras Kareiva and R. Saravanan, *Journal of Materials Science: Materials in Electronics,* Vol. 27(2), pp. 1920–1928, 2016.

Chapter 1

Introduction

Abstract

Chapter I summarizes the basic introduction to linear optics, and the physical origin of non-linear optical phenomena. It provides and discusses the principles of the characterization techniques such as, powder X-ray diffraction (PXRD), UV-Visible spectroscopy, scanning electron microscopy (SEM), and energy dispersive X-ray spectroscopy (EDS) which have been used to analyze the structural properties of the synthesized crystal systems, optical properties, morphological properties, and the elemental composition of the prepared non-linear optical materials respectively. This chapter also deals with the methodology of the structural refinement, the analysis of electron density distribution using currently available novel techniques with the corresponding computational procedure in a detailed manner, and the real space analysis of the nearest neighbor distance using atomic correlation functions. A review of the earlier work done on the chosen non-linear optical materials, $PbMoO_4$, $LiNbO_3$, $Ce:Gd_3Ga_5O_{12}$, $CaCO_3$, $Yb:CaF_2$, and Al_2O_3, $Cr:Al_2O_3$,$V:Al_2O_3$ have also been presented in this chapter. It also provides the scope of the present research work.

Keywords

NLO Materials, Physical Characterization, Powder XRD, Microstructure, Elemental Composition, SEM

Contents

1.1 Objectives

The main objective of the present research work is to study the structural properties, elucidation of the electron density distribution, optical properties, morphological properties, and the elemental composition of some non-linear optical materials. To analyse the samples, some characterization works, such as, powder X-ray diffraction (PXRD), UV-Visible spectroscopy, scanning electron microscopy (SEM) and energy dispersive X-ray spectroscopy (EDS) have been carried out.

1.2 Optics: An Introduction

Optics is an elegant and primordial topic related to the generation, propagation, detection of light, and interaction of light with matter. The conventional optics is the propagation and interaction with matter of the light from ordinary light sources, wherein the intensities of the light beam are so low that even a simple linear approximation is enough to give a good theoretical explanation for the related optical effects. The optical parameters of the medium are independent of the intensity of the light propagating in these medium. In this sense, the conventional optics may also be called linear optics.

On the other hand, the interaction of intense light (eg. laser radiation) with matter is called non-linear optics. The intensities of laser beam can be so high that a great number of new effects can be observed, and some high order non-linear approximations have to be employed to explain these new effects. The propagation of a wave through a material produces changes in the spatial and temporal distribution of electrical charges as the electrons and atoms interact with the electromagnetic fields of the wave. In this sense, non-linear optics may also be called as optics of intense light. Schrodinger (1942; 1943) introduced the term non-linear optics in 1943. The basic theories of linear and non-linear optical phenomena are discussed briefly in the next sections, though no optical studies have been made in this research work.

1.3 Linear optical medium

Light is considered as a stream of photon which does not have any rest mass, according to the quantum theory. On the other hand, light appears to be a continuous electromagnetic wave. A plane wave propagating in the direction of the vector $\vec{K}$, may be written in the form

$$\vec{E} = \vec{E_0} e^{i\,(\omega t - \vec{K}.\vec{r})} \tag{1.1}$$

where: $\vec{E}$ is the amplitude of the electric field; $\overrightarrow{E_0}$ is a constant vector; $\vec{r}$ is the position vector which defines a point in three-dimensional space; t is the time of propagation of the wave; ω is the angular frequency and $\vec{K}$ $(= \vec{i}k_x + \vec{j}k_y + \vec{k}k_z)$ is the wave vector; k_x, k_y and k_z are the components of the wave vectors along three different directions and $\vec{i}$, $\vec{j}$ and $\vec{k}$ are the unit vectors along three different directions.

When a light wave interacts with an isotropic medium, the refractive index of the medium and the velocity of the beam passing through the medium is the same in all directions. The components of the wave vectors are the same in all directions. The light beam experiences different refractive indices in different directions and propagates at different velocities in different directions in an anisotropic medium. Therefore, the components of $\vec{K}$ will have different values in different directions. The momentum $(\vec{P})$ and wavelength (λ) of a photon are related through the following expression;

$$\vec{P} = h/\lambda = \hbar\vec{K} \tag{1.2}$$

where: $\hbar$ $(= h/2\pi)$; h is the Plank's constant and $\vec{K} (= 2\pi/\lambda)$ is the wave vector. Figure 1.1 shows the schematic diagram of an optical medium without the application of an electric field.

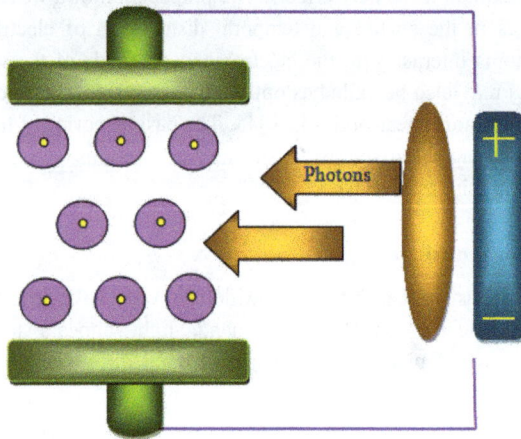

Figure 1.1 Optical medium without the application of an electric field.

4

When the optical medium is placed in an electric field, electric charges do not flow through the medium as they do in a conductor, but only slightly shift from their average equilibrium positions causing dielectric polarization. If the medium is composed of weakly bonded molecules, those molecules not only become polarized, but also reorient, so that their symmetry axis aligns to the field as shown in figure 1.2. The induced molecular dipole moment per molecule $(\vec{P})$ is proportional to the applied electric field $(\vec{E})$ which is given by the equation

$$\vec{P} = \alpha\vec{E} \tag{1.3}$$

where α is the polarizability.

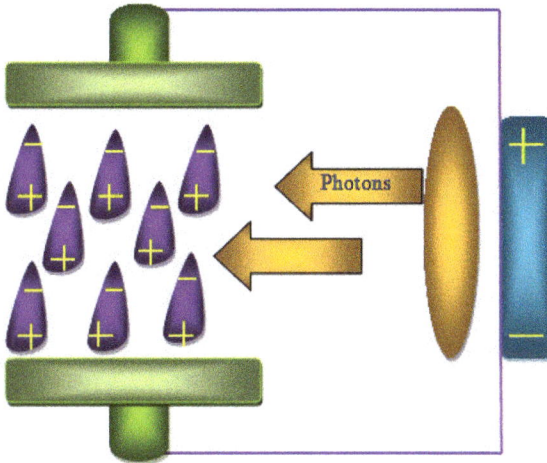

Figure 1.2 Optical medium with the application of electric field.

In a homogeneous, linear, isotropic optical medium, the response is always proportional to the stimulus. The induced polarization is proportional to the electric field and the susceptibility is independent of the electric field (Zernike and Midwinter, 1973), which is given by the equation

$$\vec{P} = \varepsilon_0 \chi \, \vec{E} \tag{1.4}$$

where ε_0 is the permittivity of free space and χ is a dimensionless proportionality constant known as the optical susceptibility of the medium. In practice, this is always the case at low electric fields. Materials in which such kind of linear relationship holds, are called linear optical medium. The variation of induced polarization ($\vec{P}$) with the applied electric field ($\vec{E}$) is a straight line as shown in figure 1.3.

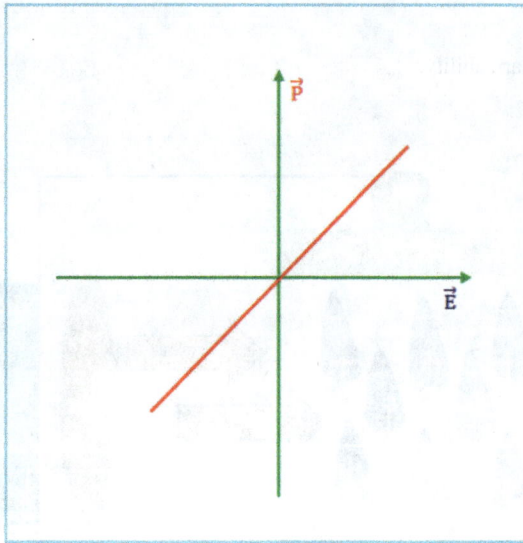

Figure 1.3 Variation of dielectric optical polarization with electric field in a linear optical medium.

The refractive index (μ) of a medium and the linear susceptibility (χ) is given by the relation

$$\mu = \sqrt{1 + 4\pi\chi} \tag{1.5}$$

As long as the intensity of light propagating in the optical medium is small, the parameters χ and μ are constants and are independent of the intensity of light. The reason is that, ordinary light sources generate light of field strengths of 10^3 V/cm to 10^5 V/cm. The field strengths are very small, compared to the atomic fields, and therefore cannot affect the optical parameters of the medium. The principle of superposition holds true in this regime. Light waves can pass through medium or be reflected from boundaries with the same frequency. The schematic representation of emitted light wave is shown in figure 1.4.

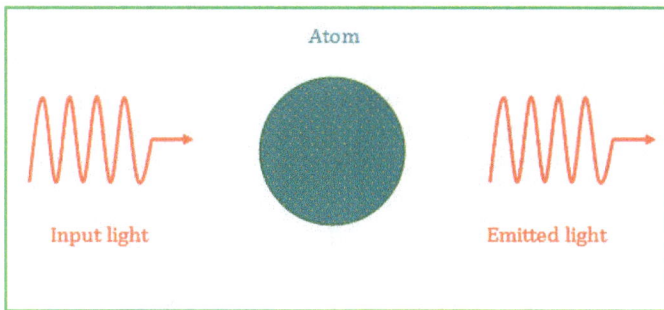

Figure 1.4 Light wave emission in a linear optical medium.

1.4 Theory of non-linear optics

It was thought that all optical media were linear throughout the long history of optics and indeed until relatively recent. The consequences of this assumption are far-reaching:

- The optical properties of medium, such as refractive index and absorption coefficient are independent of light intensity

- The principle of superposition is valid in a linear optical medium

- The frequency of light is never altered by its passage through a medium

- Two beams of light in the same region of a medium have no effect on each other so that light cannot be used to control light

Experiments carried out in the post-laser era clearly demonstrate that optical media do in fact exhibit non-linear behavior, as exemplified by the following observations:

- The refractive index and consequently the speed of light in a non-linear optical medium depends on the intensity of light
- The principle of superposition is violated in a non-linear optical medium
- Photons do interact within the confines of a non-linear optical medium so that light can indeed be used to control light
- The frequency of light is altered by its passage through a medium

The interaction of light waves can result in the generation of optical fields at new frequencies, including optical harmonics of incident radiation or sum- or difference-frequency signals. Franken and his coworkers (Franken et al., 1961), who first observed second harmonic generation by focusing the 694.3 nm output from a ruby laser into a quartz crystal and obtained a very low intensity output at a wavelength of 347.15 nm. This phenomenon with second harmonic generation and the invention of highly intense, directional and coherent laser beam has opened up new perspectives in the field of non-linear optics.

The Maxwell equations not only serve to identify and classify non-linear phenomena in terms of the relevant non-linear optical susceptibilities, but also govern the non-linear optical propagation effects. This introduction to non-linear optics is therefore limited to a simple analysis of Maxwell's equations which govern the propagation of light.

The differential form of Maxwell's equations in the presence of free charges in a dielectric media are given as

$$\nabla \cdot \vec{B} = 0 \tag{1.6}$$

$$\nabla \cdot \vec{D} = \rho \tag{1.7}$$

$$\nabla^\times \times \vec{E} = -\frac{\delta \vec{B}}{\delta t} \tag{1.8}$$

$$\nabla \times^\times \vec{H} = \vec{J} + \frac{\delta \vec{D}}{\delta t} \tag{1.9}$$

where $\vec{E}$, $\vec{B}$, $\vec{H}$, $\vec{D}$, $\vec{J}$, and ρ are the electric field, magnetic field, magnetic field strength, electric displacement, current density, and charge density respectively. Here, $\vec{B} = \mu_0 \vec{H} = \vec{E} + 4\pi\vec{M}$, where $\vec{M}$ is the magnetic dipole polarization, μ_0 is the permeability of free space and

$$\vec{D} = \varepsilon_0 \vec{E} + \vec{P} \qquad\qquad (1.10)$$

where ε_0 and $\vec{P}$ are the permittivity of free space and the induced polarization of the medium respectively. In a non-linear medium the induced polarization is a non-linear function of the applied electric field. When the high electric field strength is used, it is expected that $\vec{P}$ cannot increase linearly indefinitely with $\vec{E}$ and will become saturated. Therefore, we may anticipate non-linear behavior of $\vec{P}$ at very high field strengths. Figure 1.5 shows the non-linear variation of electric polarization with the electric field strength in a non-linear medium.

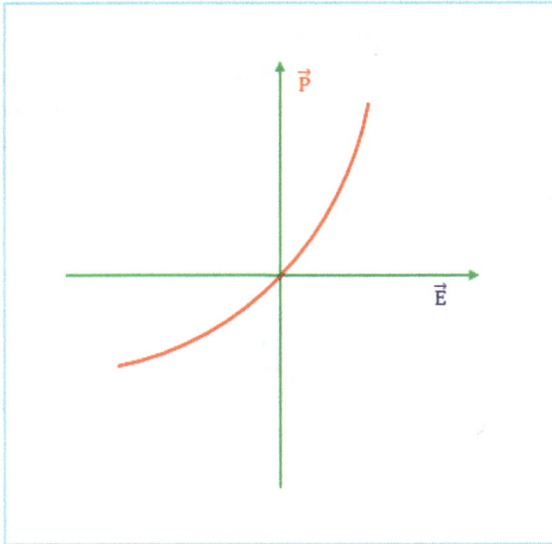

Figure 1.5 Variation of dielectric optical polarization with electric field in a non-linear optical medium.

Lasers produce light of field strengths of 10^7 V/cm to 10^{11} V/cm, which is of the order of the atomic field strength. Therefore, the intense light of lasers is in a position to cause non-linearity of $\vec{P}$ and influence the optical parameters χ, ε and μ as the function of $\vec{E}$. The superposition principle does not hold true in this regime and light waves can pass through the medium and interacting with each other and hence the frequency of the emitted wave is changed. Figure 1.6 shows the schematic representation of emitted light wave. The researchers realized that all the new effects could be reasonably explained if one replaces the linear term on the right hand side of equation (1.4) by an expansion in powers of the applied field, as (Shen, 1984)

$$\vec{P} = \varepsilon_0 \chi^{(1)} \vec{E} + \varepsilon_0 \chi^{(2)} \vec{E}^2 + \varepsilon_0 \chi^{(3)} \vec{E}^3 + \cdots \tag{1.11}$$

where $\chi^{(2)}$, $\chi^{(3)}$, ... are the second, third order, ... non-linear susceptibilities of the medium respectively. $\chi^{(1)}$ is the linear susceptibility responsible for material's linear properties like refractive index, dispersion, birefringence and absorption. $\chi^{(2)}$ is the quadratic term which describes second harmonic generation (*Franken* et al., 1961), parametric oscillation (Giordmaine and Miller, 1965) in non-centrosymmetric materials. $\chi^{(3)}$ is the cubic term responsible for third harmonic generation (Terhune et al., 1962), phase matching, and stimulated Raman scattering (Eckhardt et al., 1962). Hence the induced polarization is capable of multiplying the fundamental frequency to second, third, and even higher harmonics. In non-linear terms, product of two or more oscillating fields gives oscillation at combination of frequencies.

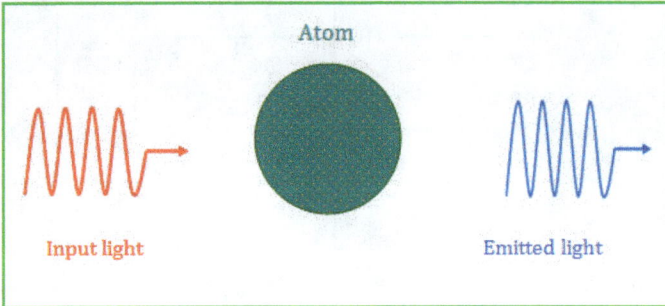

Figure 1.6 Light wave emission in a non-linear optical medium.

The induced polarization $\vec{P}$, can also be written as

$$\vec{P} = \vec{P}_l + \vec{P}_{nl} \tag{1.12}$$

where $\vec{P}_l$ is the part of the linear electric dipole polarization in the field amplitude is

$$\vec{P}_l = \varepsilon_0 \chi^{(1)} \vec{E} = \varepsilon_0 \chi_{ij} \vec{E} \tag{1.13}$$

The non-linear polarization $\vec{P}_{nl}$, which can be written as

$$\vec{P}_{nl} = \varepsilon_0 \left[\chi^{(2)} : \vec{E}\vec{E} + \chi^{(3)} : \vec{E}\vec{E}\vec{E} + \cdots \right] \tag{1.14}$$

$$\vec{P} = \vec{P}^{(1)} + \vec{P}^{(2)} + \vec{P}^{(3)} + \cdots \tag{1.15}$$

There are other non-linear optical effects, studied by numerous researchers, such as second harmonic generation, sum and difference frequency generation, phase matching, parametric oscillation, self-focusing of light and stimulated Raman scattering are available in literature. The present research work aims at the characterization of some non-linear optical materials and not to make optical studies of the materials.

The brief introduction about the various characterization techniques such as X-ray diffraction (XRD), micro-structural characterization such as scanning electron microscope (SEM), and energy dispersive analysis of X-ray (EDS) and UV-Visible spectroscopy carried out on the non-linear optical materials studied in this research work are elaborated in the next section.

1.5   Characterization techniques

1.5.1   X-ray diffraction

The structural properties of crystalline solids are determined by the symmetry of the crystalline lattice. Non-crystalline materials have no long-range order, but at least their optical properties are similar to that of crystalline materials because the wavelength of the incident photons (of the order of 1 μm) is much larger than the lattice constant of crystals. Other properties of non-crystalline materials are derived based on concepts proper to crystalline solids and, therefore, the crystal structure is extremely important in understanding the properties of solid state materials. The structural information of crystalline materials that results in this way is confirmed by X-ray diffraction experiments.

X-ray diffraction is a tool used for identifying the atomic and molecular structure of a crystal. Knowledge of the crystal structure is of crucial relevance for a proper understanding of the physical and chemical properties of the material under investigation.

X-ray crystallography relies on electromagnetic wave which interacts with matter and is used to discover the information about the structure of crystalline materials. The interaction between the incident rays with the material produces constructive interference when the Bragg's law is satisfied.

$$2d \sin\theta = n\lambda \qquad (1.16)$$

where; n is an integer, λ is the wavelength of X-rays, d is the inter-planer spacing generating diffraction and θ is diffraction angle. The Bragg's law relates the wavelength of electromagnetic wave with diffraction angle and the lattice spacing in a crystalline sample.

1.5.2 Grain size analysis from XRD

X-ray diffraction is a convenient method for determining the grain size of the polycrystalline material and the properties of the material rely on the grain size, for example, the decrease in the grain size increases the hardness of a metal or alloy (Cullity and Stock, 2001). Crystallite size can also cause peak broadening. The well known Scherrer (1918) equation explains peak broadening in terms of incident beam divergence which makes it possible to satisfy the Bragg condition for non-adjacent diffraction planes.

The software GRAIN (Saravanan, Private communication) was used to estimate approximate grain sizes from XRD. The grain size is analyzed using full width at half maximum (FWHM) of the powder XRD peaks. The particle size has been calculated using the Scherrer (1918) formula given by the equation

$$\tau = \frac{k\lambda}{\beta \cos\theta} \qquad (1.17)$$

where k is a constant related to crystallite shape, λ is the x-ray wavelength in nanometer (nm), β is the peak width of the diffraction peak profile at half maximum height resulting from small crystallite size in radians, and θ is the Bragg angle (Patterson, 1939), τ is the mean size of the ordered (crystalline) domains, which may be smaller or equal to the grain size. The dimensionless shape factor has a typical value of about 0.9, but varies with the actual shape of the crystallite. The Scherrer (1918) equation predicts crystallite thickness if crystals are smaller than 100 nm, which precludes those observed in most metallographic and ceramographic microstructures.

Phase identification using X-ray diffraction relies mainly on the positions of the peaks in a diffraction profile and to some extent on the relative intensities of these peaks. The shapes of the peaks, however, contain additional and often valuable information. The shape, particularly the width, of the peak is a measure of the amplitude of thermal oscillations of the atoms at their regular lattice sites. If all of these other contributions to the peak width is zero, then the peak width would be determined solely by the particle size and the Scherrer (1918) formula would apply. If the other contributions to the width are non-zero, then the particle size can be larger than that predicted by the Scherrer (1918) formula, with the extra peak width coming from the other factors. It is worthy to note that the Scherrer (1918) formula provides a lower bound on the particle size. In the present investigation, the grain size has been analyzed from XRD. The grain morphology was examined by scanning electron microscopy (SEM).

1.5.3 Micro-structural characterization

Micro-structural characterization has become important for all types of materials as it gives substantial information about the structure-property correlation. Micro-structural characterization broadly means ascertaining the morphology, identification of crystallographic defects and composition of phases, estimating the particle size, etc. Electron microscopic techniques are extensively used for this purpose. Electron microscopy is based on the interaction between electrons (matter wave) and the sample. The electrons interact with atoms in the sample, producing various signals and they contain the information about surface morphology and compositions. The principle and experimental details of scanning electron microscope (SEM) analysis and energy dispersive X-rays spectroscopy (EDS) are given below.

1.5.4 Scanning electron microscopy

Scanning electron microscope (SEM) is one of the best tools for the morphological studies. Scanning electron microscope produces an image of the sample by scanning it with the focused beam of electrons. In a typical scanning electron microscope, a well-focused electron beam (source of illumination) is made to fall and scan the sample surface by two pairs of electro-magnetic deflection coils. The electron beam usually has an energy ranging from 0.2 KeV to 40 KeV which is focused by one or two condenser lenses to a spot about 0.4 nm to 5 nm in diameter. The ejected electron beams are directed towards the sample through the lenses. The lenses are functioning as optical microscope lenses which produce clear and detailed image. These lenses are made up of magnets capable of bending the path of electrons and they focus and control the electron beam. The beam passes through a final lens which deflects the beam along x and y axes. The sample chamber keeps the samples and it is very sensitive to vibrations. The deflected electron beam is detected by the detector. For carrying out SEM analysis, the sample must be vacuum compatible (~10^{-6} Torr or more) and electrically conducting. The surfaces of non-conductive materials are made conductive by coating with a thin film of gold or platinum or carbon. The schematic diagram of the scanning electron microscope (Atteberry, 2014) is shown in figure 1.7.

The SEM is also capable of performing analyses of selected point locations on the sample. SEM micrographs have a large depth of field yielding a characteristic three-dimensional appearance useful for understanding the surface structure of a sample. Areas ranging from approximately 1 cm to 5 microns in width can be imaged in a scanning mode using conventional SEM techniques (magnification ranging from 20X to approximately 100000X, spatial resolution of 50 to 100 nm).

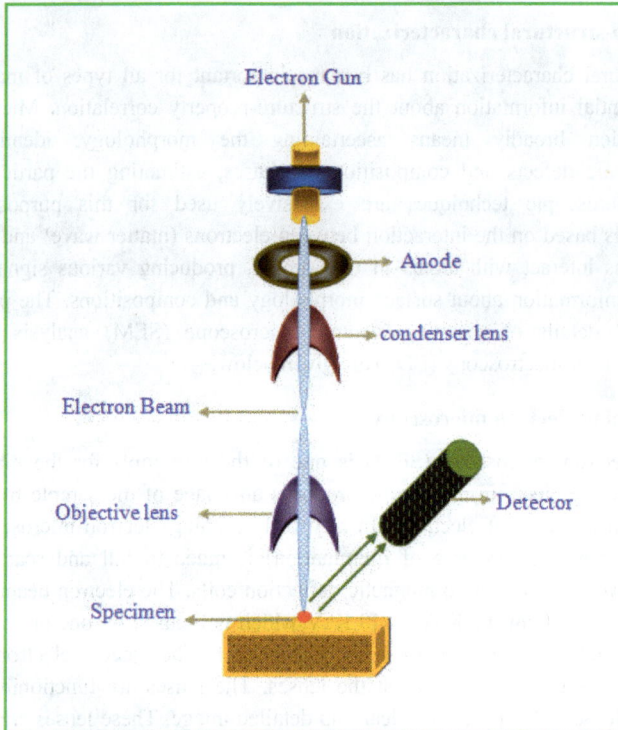

Figure 1.7 Schematic diagram of scanning electron microscope.

1.5.5 Energy Dispersive X-ray spectroscopy (EDS)

The energy dispersive X-ray spectroscopy (EDS) is an analytical technique used for qualitative analysis (the identification of the lines in the X-ray spectrum) and quantitative analysis (determination of the concentrations of the elements present) involves measuring line intensities for each element in the sample. The schematic diagram of the EDS arrangement is shown in figure 1.8.

The electron beam stimulates the atoms in the sample with uniform energy and they instantaneously drive out characteristic X-rays of specific energies for each element. An EDS X-ray detector is used to separate the characteristic X-rays of different elements into an energy spectrum and EDS system software is used to analyze the energy spectrum in order to determine the abundance of specific elements. The most common detectors are made of Si (Li) crystals that operate at low voltages to improve sensitivity. In recent years, advances in detector technology (silicon drift detectors) operate at higher count rate without liquid nitrogen cooling.

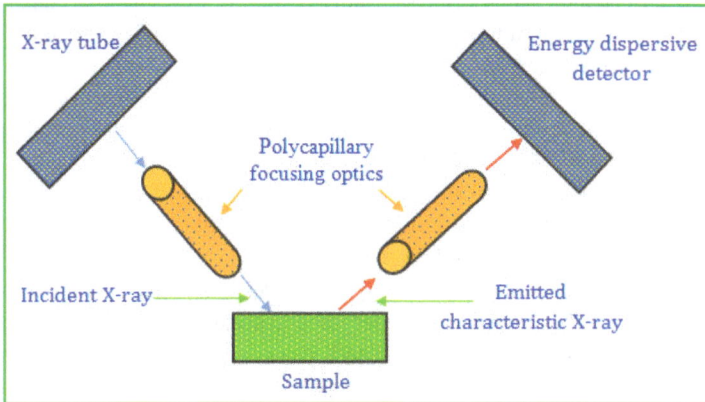

Figure 1.8 Schematic diagram of EDS arrangement.

A typical EDS spectrum is portrayed as a plot of X-ray counts vs. energy (in keV). Energy peaks correspond to the various elements in the sample. EDS can be used to find the chemical composition of materials down to a spot size of a few microns and to create element composition maps over a much broader raster area. Together, these capabilities provide fundamental compositional information for a wide variety of materials. This radiation gives information about the elemental composition of the sample.

1.5.6 UV-Visible spectroscopy

The Ultraviolet-Visible spectroscopy is a technique involving the use of light in the visible (380-780 nm), near ultra-violet (200-380 nm) and near infrared regions (780 - 1100 nm) to cause electronic transitions in the target material. A light source of a fixed wavelength is shone through a sample and its absorption (or transmission) intensity is measured against a background using a detector. The wavelength is then varied slightly using a diffractometer, and the process is repeated until the absorption ratio for a spectrum of wavelengths is obtained.

UV-Visible spectroscopy is an important tool in analytical research. The indispensable elements of a UV-Visible spectrophotometer (Gullapalli and Barron, 2010) consist of a light source, a spectrometer, a sample compartment, and a detector. The most commonly used light sources are the halogen lamps used for the visible and near-infrared regions and the deuterium lamps used for the ultraviolet region. Dispersion devices cause different wavelengths of light to be dispersed at different angles. A monochromator consists of an entrance slit, a dispersion device, and an exit slit. The output from a monochromator is usually a monochromatic light. When combined with an appropriate exit slit, these devices can be used to select a particular wavelength of light from a continuous source. Two cells are required for the split-beam instrument. The monochromatic light divided by beam splitter is then passed through the reference cell and sample cell. Both cells are initially filled with pure solvent, and a so-called balance measurement is performed. This measurement reflects the difference in absorbance between the two optical paths in use. The sample cell is then filled with the sample solution for measurement. The resulting spectrum is corrected by subtracting the balance spectrum. A detector converts a light signal into an electrical signal. Spectrophotometers normally contain either a photomultiplier tube detector or a photodiode detector. Increasingly, photodiodes are used as detectors in spectrophotometers. Impurities can be detected very easily by testing if the compound is not showing its characteristic absorption spectrum. The molecular weight of the compound is calculated on the basis of absorption of data. The band gap of the crystalline sample can be determined. The schematic diagram of UV-Visible spectrophotometer is as shown in figure 1.9.

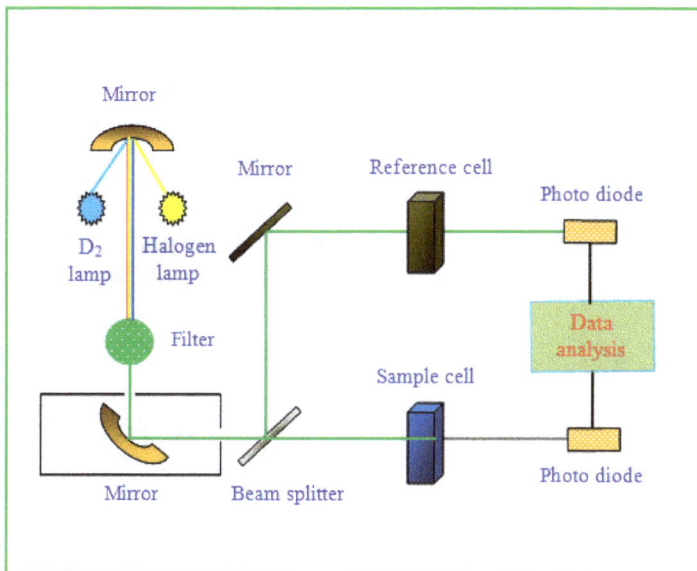

Figure 1.9 Schematic diagram of UV-Visible spectrophotometer.

1.6 Theoretical framework on charge density

1.6.1 Structure factors and electron density

The examination of X-ray diffraction pattern provides the details of the dimension and the shape of the unit cell. One can calculate the structure if the molecule consists of a single scatterer at the origin of the unit cell. The sum of the wavelets scattered from all the infinitesimal elements of electron density in the unit cell is the structure factor. The structure factor, F_{hkl}, is the resultant of j waves scattered in the direction of the reflection $(h\ k\ l)$ by the j atoms in the unit cell. Each of these waves has amplitude proportional to f_j, the scattering factor of the atom.

The structure factor in exponential form is given by the relation,

$$F(\vec{S}) = F_{hkl} = \sum_{j=1}^{N} f_j\, exp2\pi i\left(hx_j + ky_j + lz_j\right) \qquad (1.18)$$

17

where N is the number of atoms in the unit cell. This means that X-ray scattering is dependent on electron density ρ, the number of electrons per unit volume. In terms of electron density, the structure factor for the volume element is given by the equation

$$F_{hkl} = \int_V \rho(X,Y,Z) exp2\pi i\left(hx_j + ky_j + lz_j\right)dV \tag{1.19}$$

where $\rho(X,Y,Z)$ is the number of electrons per unit volume. The electron density with the help of Fourier transform can be expressed in terms of structure factor as

$$\rho(X,Y,Z) = \frac{1}{V}\sum\sum\sum F_{hkl}exp[-2\pi i\left(hx_j + ky_j + lz_j\right)] \tag{1.20}$$

The three dimensional electron density equation for the two dimensional calculation reduces to

$$\rho(X,Y,Z) = \frac{1}{V}\sum\sum F_{hkl}exp[-2\pi i\left(hx_j + ky_j\right)] \tag{1.21}$$

One can generate an electron density map for the unit cell of the crystal if the electron densities calculated from equation (1.21) are displayed for the (x, y, z) positions in the unit cell.

All atoms and the related electrons contribute to the structure factor of each observed reflection, and all observed reflections must therefore be used to determine the types and positions of each atom in space which can easily be revealed from equations (1.18) and (1.19). We can truncate the Fourier series at some finite values of $(h\ k\ l)$ in equation (1.21) because it is not possible to measure all possible reflections. It means that a completely accurate description of the contents of the unit cell is not possible. But, the electron density at any point in space and the structure of the crystal can readily be calculated if the structure factors for a large number of scattering vectors $(\vec{S})$ are included. The scattering vectors $(\vec{S})$ cannot be directly measured in a diffraction experiment but it can be calculated from a known set of atomic positions. The intensity of the diffracted X-ray for each reflection and the structure factors related to the intensity can be easily measured but, some of the critical information may be lost in the intensities that prevent directly solving the atomic structure of a crystal from the observed X-ray diffraction data.

1.6.2 Least-squares refinement

Let us imagine an error-free set of $|F_o|$ and an almost correct set of atomic co-ordinates and temperature factors. For simplicity of our discussion, one can assume that the structure is centro-symmetric and the temperature factors are isotropic. The calculated structure factors is given by the equation,

$$(F_o)_{hkl} = \sum_{j=1}^{N/2} 2f_i \, cos \, 2\pi(h\vec{x_j} + k\vec{y_j} + l\vec{z_j}) \, [-B_j(\tfrac{sin^2 \theta}{\lambda^2})] \tag{1.22}$$

The correct thermal and positional parameters i.e., B_j, $\vec{x_j}$, $\vec{y_j}$ and $\vec{z_j}$ for the jth atom can be expressed as $B_j + \Delta B_j$, $\vec{x_j} + \Delta x_j, \vec{y_j} + \Delta y_j$ and $\vec{z_j} + \Delta z_j$ and hence one can write equation (1.22) as

$$(F_o)_{hkl} = \sum_{j=1}^{\frac{N}{2}} 2f_i \, cos \, 2\pi[h(\vec{x_j} + \Delta\vec{x_j}) + k(\vec{y_j} + \Delta\vec{y_j}) + l(\vec{z_j} + \Delta\vec{z_j})] \, *$$

$$[(-B_j + \Delta B_j)(\tfrac{sin^2 \theta}{\lambda^2})] \tag{1.23}$$

N equivalent reflections of a given Laue class are separated into two subsets of $(x/2)$ equivalent reflections in the equation (1.23), then,

$$\Delta F_{hkl} = \sum_{j=1}^{N/2} \frac{\partial (F_C)_{hkl}}{\partial B_j} B_j + \frac{\partial (F_C)_{hkl}}{\partial x_j} \Delta\vec{x_j} + \frac{\partial (F_C)_{hkl}}{\partial y_j} \Delta\vec{y_j} + \frac{\partial (F_C)_{hkl}}{\partial z_j} \Delta\vec{z_j} \tag{1.24}$$

where,

$$\Delta F_{hkl} = (F_o)_{hkl} - (F_c)_{hkl} \tag{1.25}$$

Equations of type (1.24) and (1.25) can be produced for each reflection and in general, there will be many more equations than parameters. If a subset of these equations was taken such that the number of equations equaled the number of parameters, then the correct parameters could be found and recalculated. Values of F_o's would then equal to the $|F_o|$'s. However in practice, the $|F_o|$'s are not error free and so one finds a least–square solutions of the complete set of equations. This solutions is the one such that when the parameters are changed to $B_j + \Delta B_j$, $\vec{x_j} + \Delta\vec{x_j}, \vec{y_j} + \Delta\vec{y_j}$ and $\vec{z_j} + \Delta\vec{z_j}$ the quantity,

$$R_S = [(F_o)_{hkl} - (F_c)_{hkl}] \tag{1.26}$$

is a minimum, where $(F_c')_{hkl}$ is the revised calculated structure factor. From Taylor's theorem, one can find, ignoring second order small quantities, then,

$$(F_c')_{hkl} = (F_c)_{hkl} - \left(\sum_{j=1}^{N/2} \frac{\partial (F_C)_{hkl}}{\partial B_j} B_j + \frac{\partial (F_C)_{hkl}}{\partial x_j} \Delta \vec{x_j} + \frac{\partial (F_C)_{hkl}}{\partial y_j} \Delta \vec{y_j} + \frac{\partial (F_C)_{hkl}}{\partial z_j} \Delta \vec{z_j}\right) \tag{1.27}$$

$$(F_o)_{hkl} = (F_c')_{hkl} - \left(\sum_{j=1}^{N/2} \frac{\partial (F_C)_{hkl}}{\partial B_j} B_j + \frac{\partial (F_C)_{hkl}}{\partial x_j} \Delta \vec{x_j} + \frac{\partial (F_C)_{hkl}}{\partial y_j} \Delta \vec{y_j} + \frac{\partial (F_C)_{hkl}}{\partial z_j} \Delta \vec{z_j}\right) \tag{1.28}$$

In equation (1.26), R_S is the sum of the squares of the difference between the left-hand sides and right-hand sides of equations (1.27) and (1.28) and it is the minimization of this quantity which is usually implied by the least squares solution.

It is possible to weight the equations according to the expected reliability of the quantity ΔF_{hkl}. If the measured $|F_o|$ expected to have a large random error, then the value of ΔF_{hkl} may well be dominated by this and ΔF_{hkl} may not be very useful in indicating the changes to be made in the atomic parameters. Errors of measurement tend to be related to the actual value of F_o and many weighting schemes have been proposed based on the value of $|F_o|$. When a reasonable set of weights has been found, the equation for R_S is multiplied throughout by the appropriate weight and a least squares solution of the modified set of equations is then sought in the usual way. The standard least squares approximate procedure by the full matrix method is adopted for refining the parameters such as scale factors, thermal parameters and extinction parameter. In the least square procedures, the quantity to be minimized is

$$R_{sw} = \sum_{hkl} W_{hkl} (|F_o| - |kF_c|^2) \tag{1.29}$$

where in the above equation (1.29), W_{hkl} is the weight to be assigned on an observation, F_o is the observed structure factor, F_c is the calculated structure factor and k is the scale factor. From the final value of R_{sw}, the standard errors of the final parameters can be estimated. Equation (1.29) is a measure of the degree to which the distribution of difference between $|F_o|$ and $|F_c|$ fits the distribution expected from the weights used in the refinement (International Tables for X-ray Crystallography 1974).

The calculation of charge density by this least square refinement method we need infinite number of Fourier co-efficient to perform the Fourier Synthesis. But we use only a limited number of Fourier co-efficient. We neglect the missing structure factors by setting them to zero simply because the experiment cannot be carried out. This is a highly biased assumption. This results in the unphysical negative electron density and hampers the use of it in understanding the finite details like the bonding charge in valence region.

1.7 X-ray powder refinement technique through Rietveld method

The Rietveld (1969) method is a technique used for the characterization of crystalline materials from X-ray powder diffraction data. The neutron and X-ray diffraction of powder samples result in a pattern characterized by reflections (peaks) in intensity at certain positions. The success of the Rietveld (1969) method relies on accurate description of the shapes of the peaks in a powder pattern. Various aspects of the materials structure can be easily determined by the height, width and position of these reflections.

A curve fitting procedure for analyzing the more complex patterns obtained from low symmetry materials was proposed by Rietveld (1969). The difference between the observed and calculated profiles rather than individual reflections was minimized by the least squares refinement. The best fit between the experimental and calculated powder diffraction patterns is achieved by adjusting the variables defining the structural model (atomic positions and atomic displacement parameters) and the powder diffraction profile (unit cell parameters and zero shift error, analytical functions describing the peak shape and peak width, background intensity coefficients) in the Rietveld (1969) method of crystal structure determination using powder diffraction data by least-squares methods. The pseudo-Voigt function (Thompson et al., 1987), which represents a hybrid of Gaussian and Lorentzian characters of peak shape is the most widely used peak shape function for X-ray powder diffraction data. The lattice parameters, which determine the positions of the reflections, a zero point error for the detector and three parameters that describe the variation of the Gaussian half width (full width at half maximum intensity) with scattering angle are the parameters required in addition to the conventional parameters in the least squares procedure. Several authors have applied and reviewed this technique to a wide range of solid state problems (Cheetham and Taylor, 1977; Hewat, 1986; Young, 1993).

The iterative procedure was first reported for the diffraction of monochromatic neutrons where the position of reflection is reported in terms of 2θ values. The poor powder averaging (large crystallites within a sample of smaller ones) and preferred orientation, both of which arise from the fact that X-rays probe a smaller sample volume are the other problems that occur in the X-ray diffraction studies.

The field of applications of Rietveld (1969) method involved in the use of neutrons received an additional improvement in the late 1970's with the extension to X-ray data (Malmros and Thomas, 1977; Young et al., 1977). High resolution data collected at a synchrotron source leads to an accurate and precise determination of crystal structure (Cox et.al., 1983). The slight variations in peak shape from one reflection to another may be a frustration for the crystallographer, if the aim is to achieve a high quality of the crystal structure. In contrast, the additional information pertaining to the microstructure of the sample can easily be retrieved by the materials scientist. Several inherent parameters are discussed in the following sections in order to understand the Rietveld (1969) method.

1.7.1 The Rietveld strategy

A step scanning method was used for the collection of powder diffraction patterns. Bragg diffraction intensity with two-theta axes was collected for the crystal structure determination (Young, 1993). The momentum transfer Q, in all other diffraction and optics techniques but rarely used in powder diffraction with the wavelength is given by the relation

$$Q = 4\pi sin(\theta)/\lambda \tag{1.30}$$

1.7.2 Peak shape function

The instrumental arrangement, the characteristics of the X-ray beam and the nature (size and shape) of the sample influences the peak shape in a X-ray powder diffraction pattern. The calculated intensity at each point on the profile is obtained by summing the contributions from the Gaussian peaks that overlap at that point by assuming that the reflections with constant wavelength X-ray are Gaussian in shape. The contribution of a given reflection to the profile y_i at position $2\theta_i$ is given as

$$y_i = I_K \exp[-4\ln\{2\,(2\theta_i - 2\theta_K)/H_K)^2\}] \tag{1.31}$$

where H_K is the full-width half-maximum (FWHM), $2\theta_K$ is the centre of the reflection, and I_K is the calculated intensity determined from the structure factor, the Lorentz factor, and multiplicity of the reflection.

The reflections may acquire an asymmetry due to the vertical divergence of the beam, at very low diffraction angles. To account for this asymmetry, Rietveld (1969) used a semi-empirical correction factor, A_S

$$A_S = 1 - [sp(2\theta_i - 2\theta_\kappa)^2/tan\,\theta_\kappa] \tag{1.32}$$

where p is the asymmetry factor and s is -1, 0, +1 depending on the difference $(2\theta_i - 2\theta_\kappa)$ being negative, zero, or positive respectively. More than one diffraction peak may contribute to the profile at a given position. The sum of all reflections contributing at the point $2\theta_i$ is simply the diffraction intensity.

1.7.3 Peak width function

The second inherent parameter is the width of the diffraction peaks and found to be broad at higher Bragg angles. This angular dependence (Caglioti et al., 1958) of the half widths of the diffraction peaks is given as

$$\sigma^2 = H_K^2 + Psec^2\theta_\kappa = Utan^2\theta_\kappa + V\,tan\theta_\kappa + W + Psec^2\theta_\kappa \tag{1.33}$$

where U, V and W are the Gaussian FWHM adjustable parameters and the Scherrer coefficient, P, for Gaussian broadening in Rietveld's (1969) least squares calculations. These four parameters do not converge in a stable manner when refined simultaneously (Prince, 1993). The parameters V and W in equation (1.33) depend only on instruments but not on specimens (Young and Desai, 1989) and may be fixed at values obtained by the Rietveld (1969) refinement of a well-crystallized sample where profile broadening is negligible, i.e., $P = 0$. $H_{KL,}$ the Lorentzian FWHM's for the Voigt function corresponding to the pseudo-Voigt function varies with θ_K as

$$H_{KL} = (X + X_e cosj_K)sec\theta_K + (Y + Y_e cosj_K)tan\theta_K \tag{1.34}$$

The first and second parts proportional to $sec\theta_K$ has the angular dependence associated with Scherrer (1918) approximation for crystallite-size broadening and $tan\theta_K$ related to Lorentzian micro strain broadening respectively. X and Y are isotropic-broadening coefficients (Thompson et al., 1987). On the other hand, X_e and Y_e are anisotropic-broadening coefficients (Larson and Von Dreele, 1990) and j_K is the angle between the scattering vector, $Q_K(= ha^* + kb^* + lc^*)$ and an anisotropic broadening axis, $h_a a^* + k_a b^* + l_a c^*$.

The parameter P in equation (1.33) provides a component of the Gaussian FWHM which is constant in d^*, as is the X component in equation (1.34) (Young, 1993). The modification of the original pseudo-Voigt function of Thompson et al. (1987) consists of the Scherrer coefficient, P and the anisotropy coefficients, X_e and Y_e (Larson and Von Dreele, 1990). Equations that relate U, P, X and Y to the crystallite size and microstrain are described in Larson and Von Dreele (1990).

1.7.4 Profile asymmetry and peak shift

Howard (1982) devised a profile asymmetry introduced by employing a multi-term Simpson's rule integration, where n symmetric profile-shape functions with different Simpson's coefficients for weights, g_j and shifts, f_j are positioned asymmetrically and superimposed with each other

$$\phi'(\Delta 2\theta) = \frac{1}{3(n+1)} \sum_{j=1}^{n} g_j \, \phi(\Delta 2\theta) \tag{1.35}$$

with

$$\Delta 2\theta' = \Delta 2\theta + f_j A_s \cot 2\theta_k + Z + D_s \cos \theta_k + T_s \sin 2\theta_k \tag{1.36}$$

Here, $\phi'(\Delta 2\theta)$ is the asymmetric pseudo-Voigt function and $\Delta 2\theta'$ is the 2θ difference modified for (a) profile asymmetry (Howard, 1982) and (b) peak shifts for each component profile such as zero-point shift, Z, specimen displacement, D_s and specimen transparency, T_s. The corresponding Simpson's coefficients are,

$n = 3$: $g_1 = g_3 = 1$; $g_2 = 4$

$n = 5$: $g_1 = g_5 = 1$; $g_2 = g_4 = 4$; $g_3 = 2$

$n = 7$: $g_1 = g_7 = 1$; $g_2 = g_4 = g_6 = 4$; $g_3 = g_5 = 2$

$n = 9$: $g_1 = g_9 = 1$; $g_2 = g_4 = g_6 = g_8 = 4$; $g_3 = g_5 = g_7 = 2$ $\qquad$ (1.37)

$$f_j = [(j-1)/(n-1)]^2 \tag{1.38}$$

The number of terms, n (= 3, 5, 7, or 9), in equation (1.38) is automatically adjusted for each reflection using its FWHM.

This method of making the profile shape asymmetric gives better fits to asymmetric profiles than the simple one proposed at first by Rietveld (1969), showing less correlation with lattice parameters. Howard's (1982) approach further offers physical insight into the origin of the asymmetry because it is based explicitly upon axial divergence. It may, however, fail to fit strongly asymmetric profiles at very low scattering angles. In fact, the Simpson's rule integration can break up into multiple peaks for very strong asymmetry.

The asymmetric pseudo-Voigt function (Thompson et al., 1987), $\phi'(\Delta 2\theta)$, composed of equations (1.32) - (1.38) contain the nine profile-shape parameters (U, V, W, P, X, Y, X_e, Y_e and A_S) and the three peak-shift parameters (Z, D_s and T_s) that can be refined in Rietveld (1969) analysis. This function is sound in that it has a physical foundation as well as just fitting the observed diffraction pattern. It can extract microstructural information, $i.e.$, crystallite size and microstrain, from isotropic and/or anisotropic broadening of profiles. No line broadening arises from the crystallites as long as the experimental setup is unchanged.

1.7.5 Preferred orientation function

In order to obtain a representative diffraction pattern in reflection geometry, sample presentation is a key aspect. In polycrystalline specimens, crystal morphology causes preferred orientation of particles. Plate-like clay particles and fibrous materials are the typical examples of specimens affected by preferred orientation. In such cases, the intensity distribution along the Debye rings is not constant because of the random orientation of the grains. If the shape of the particle is spherical, then only an absolute random orientation of particles is possible. The information about morphological characteristics of sample can readily be used for the various correction procedures. Single variable functions describing orientation with respect to a specific oriented reflection plane of the phase (Dollase, 1986) is the most effective models used in Rietveld (1969) refinement. The use of the Rietveld (1969) method for eliminating (or minimizing) preferred orientation can be written as,

$$I_{corr} = I_{obs}\, exp(-G a^2) \tag{1.39}$$

where I_{obs} is the intensity expected for a random sample, G is the preferred orientation parameter and α is the acute angle between the scattering vector and the normal of the crystallites. A brief outline about the Rietveld (1969) refinement procedure is described in the following section.

1.7.6 Refinement procedure

The Rietveld (1969) method of determining structural parameters can begin with a complete structural model and good starting values for the background contribution, the unit-cell parameters and the profile parameters. The numerical intensity y_i at each of the several thousand equal steps along the scattering angle 2θ, with increments $\Delta 2\theta$, is the basis for the procedure. The step size may range from 0.01° to 0.05°. The basic principle of the Rietveld (1969) method is to minimize a function M which analyzes the difference between a calculated profile $y(cal)$ and the observed data $y(obs)$. The convergence criterion for this, is given as

$$M = \sum_i W_i [y_i(obs) - \frac{1}{c} y_i(cal)] = minimum \tag{1.40}$$

where W_i is the statistical weight of each observation point and c is an overall scale factor such that $y_i(cal) = cy_i(obs)$. The main parameters of this refinements are: the profile parameters, the half width parameters (U, V and W), possible asymmetries (P) of the diffraction peaks and a zero point adjustment (Z), the unit cell parameters ($a, b, c, \alpha, \beta, \gamma$), the crystallographic symmetry, especially space group and preferred orientation parameter (G), and structural parameters like the overall scale factor c, fractional positional coordinates of the jth atom in the asymmetric unit (x_i, y_i, z_i), atomic (isotropic) Debye-Waller factors (including anisotropic parameters), B_j and occupation number of each crystallographic site (N_j) with j^{th} atom in the asymmetric unit.

It is necessary to provide the approximate values of all the parameters for the first cycle. These parameters are refined until a certain convergence criterion is reached, or the refinement is stopped by the operator in subsequent refinement cycles. The background parameters of the profile are fitted to suitable 6 parameter or 12 parameter models etc., which are available in a Rietveld (1969) profile fitting methodology. The asymmetric peak widths can be refined using Bearer-Baldinozzi (asym1, asym2, asym3 and asym4) and the profile fitting model such as voigt, pseudo-voigt etc., can also be used for profile fitting. The micro absorption effects, surface roughness, temperature factors, occupancy of atoms and composition of atoms are the examples of some other parameters that can be refined.

Nowadays, fast detectors (image plate readers) in combination with powerful microcomputers reveal a new aspect of Rietveld-refinement (Rietveld, 1969). The change of crystal structure with pressure, temperature and/or during a chemical reaction is monitored and visualized by recording full powder patterns in short time intervals. The decomposition formula can be iterated in such a way that at the point where calculated structure factors are entered, a set of identical is given instead. In Rietveld (1969) refinement, the calculated set of $|F_{obs}|$ from the decomposition formula can then be used as a new $|F_{cal}|$.

Rietveld (1969) method calculates and improves the entire powder pattern using a variety of refinable parameters by minimizing the weighted sum of the squared differences between the observed and the calculated pattern using least square methods. Finally, the structure factors evolved from the Rietveld (1969) refinements were further utilized for the estimation of charge density in the unit cell. The reliability and success of this method can be gauged by the publication of more than a thousand scientific papers.

1.7.7 Rietveld strategy with JANA2006

In the present work, the Rietveld (1969) refinement of (i) $PbMoO_4$, (ii) $LiNbO_3$, (iii) $Ce:Gd_3Ga_5O_{12}$, (iv) $CaCO_3$, (v) $Yb:CaF_2$, and (vi) Al_2O_3, $Cr:Al_2O_3,V:Al2O_3$ have been performed using the software JANA 2006 (Petříček et. al., 2006). JANA2006 (Petříček et. al., 2006), is a crystallographic program for solving and refinement of regular, modulated and composite structures from both single crystal diffraction data and powder diffraction data. Structure analysis from data reduction and powder profile analysis to the solution of the phase problem, structure refinement and presentation of results are the basic tasks covered by JANA2006 (Petříček et. al., 2006). The structure solution can be done using the built-in charge flipping algorithm or by calling an external direct methods program. Multiphase structures (for both powder and single crystal data), as well as twins with partial overlap of diffraction spots, commensurate and composite structures can be easily handled by JANA2006 (Petříček et. al., 2006). Powerful transformation tools for symmetry, cell parameters and commensurate-supercell relations are comprised in JANA2006 (Petříček et. al., 2006). Finally, the structure factors evolved from the Rietveld (1969) refinements are given in appropriate sections, and then the refined structure factors have been utilized for maximum entropy method (MEM) refinements to explicate the electron density distribution, which will be discussed in the next section.

1.8 Entropy maximized charge density distribution

A versatile approach to the estimation of spatial electron density distribution in a solid crystal is the maximum entropy method (MEM). The maximum entropy method (MEM) is a method to derive the most probable map based on a non-linear calculation by the use of information for a given set of experimental diffraction data. A precise electron-density map can be obtained and the existence of bonding between the atoms is clearly visible in the maximum entropy map.

Collins (1982) introduced the improved statistical approach, maximum entropy method (MEM) to deal various crystallographic problems. Takata et al. (2001) have applied it actively to the determination of electron densities, using synchrotron X-ray diffraction (XRD) data. The review of Gilmore (1996) also established that the MEM is most suited in the reconstruction of charge density. MEM reports have provided a positive impetus for using MEM analysis to study charge density, ion conduction, and atomic disorder (Itoh et al., 2010).

The steps for structure modeling and imaging of charge density are based on the preliminary reference model from Rietveld (1969) refinement. A rigid body model and restraints for intermolecular distances and angles were applied in the refinement. Using the result of Rietveld (1969) refinement, the integrated intensities of each reflection are evaluated from the observed diffraction data. After several iterative refinements, the final charge density was obtained.

1.8.1 Overview of theoretical explanation of MEM

An information-theory-based technique, MEM (Collins, 1982) was first developed in the field or radio astronomy to enhance the information obtained from noisy data (Gull and Daniel, 1978). It is based on the theory of foundation of statistical thermodynamics. Both the statistical entropy and the information entropy deal with the most probable distribution.

The probability of a distribution of N identical particles over m boxes, each populated by n_i particles, is given by

$$P = \frac{N!}{n_1! n_2! n_3! ... n_m!} \tag{1.41}$$

As in statistical thermodynamics, the entropy is defined as $\ln(P)$. Since the numerator is constant, the entropy is, apart from a constant, equal to

$$S = -\sum_i n_i \, ln(n_i) \tag{1.42}$$

where Stirling's formula $(ln\,N! \approx N\,ln\,N - N)$ has been used.

In case, there is a prior probability q_i for box i to contain n_i particles, equation (1.41) becomes

$$P = \frac{N!}{n_1! n_2! n_3! \ldots n_m!} \, q_1^{n_1} q_2^{n_2} q_3^{n_3} \ldots q_m^{n_m} \tag{1.43}$$

which gives, for the entropy equation,

$$S = -\sum_i n_i \, ln(n_i) + \sum_i n_i \, ln(q_i) = -\sum_{i=1}^{m} n_i \, ln\left(\frac{n_i}{q_i}\right) \tag{1.44}$$

The maximum entropy method was first introduced into crystallography by Collins (1982), who expressed the information based on equation (1.44), the entropy of the electron density distribution as a sum over m grid points in the unit cell, using the entropy formula (Jaynes, 1968)

$$S - \sum \rho'(r) \, ln\left[\frac{\rho'(r)}{\tau'(r)}\right] \tag{1.45}$$

where the probability $\rho'(r)$ and prior probability $\tau'(r)$ are related to the actual electron density in a unit cell as,

$$\rho'(r) = \frac{\rho(r)}{\sum_i \rho(r)} \tag{1.46}$$

and

$$\tau'(r) = \frac{\tau(r)}{\sum_i \tau(r)} \tag{1.47}$$

where $\rho(r)$ and $\tau(r)$ are the electron density and prior electron density at a certain fixed r in a unit cell, respectively. In the present theory, the actual densities are treated hereafter instead of normalized densities. And $\rho'(r)$ becomes $\tau'(r)$ when there is no information. The $\rho'(r)$ and $\tau'(r)$ are normalized as

$$\sum \rho'(r) = 1 \ and \ \sum \tau'(r) = 1 \tag{1.48}$$

A constraint is introduced here as,

$$C = \frac{1}{N} \sum_k \left[\frac{|F_{cal}(k) - F_{obs}(k)|^2}{\sigma^2(k)}\right] \tag{1.49}$$

Where N is the number of reflections used for MEM analysis, $\sigma(H)$ is standard deviation of $F_{obs}(H)$, the observed structure factor and $F_{cal}(H)$ is the calculated structure factor given by,

$$F_{cal}(H) = V \sum_r \rho(r) \, exp(-2\pi i H.r) \, dV \tag{1.50}$$

where V is the unit cell volume.

This type of constraint is sometimes called a weak constraint, in which the calculated structure factors agree with the observed ones as a whole, when C becomes unity. As can be seen in equation (1.50) the structure factors are given by the Fourier transform of the electron density distribution in a unit cell. In the MEM analysis, there is no need to introduce the atomic form factors, by which the structure factors are normally written. It should be emphasized here that it would be an assumption to use the atomic form factors in the formulation of the structure factors. Equation (1.50), guarantees that it is possible to allow any kind of deformation of the electron densities in real space as long as information concerning such a deformation is included in the observed data.

We use Lagrange's method of undetermined multiplier (λ) in order to constrain the function C to be unity while maximizing the entropy.

We then have

$$Q = S - \left(\frac{\lambda}{2}\right) =$$

$$-\sum \rho'(r) \, ln\left[\frac{\rho'(r)}{\tau'(r)}\right] - \frac{1}{2N} \sum_k \left[\frac{|F_{cal}(k) - F_{obs}(k)|^2}{\sigma^2(k)}\right] \tag{1.51}$$

and when $dQ/d\rho = 0$ and using the approximation, $ln\,x = x - 1$ we get,

$$\rho(r_i) = \tau(r_i) exp\left\{\left(\frac{\lambda F_{000}}{N}\right)\left[\sum_k \frac{(F_{obs}(k) - F_{cal}(k))}{\sigma(k)^2}\right] exp(-2\pi i k.r)\right\} \tag{1.52}$$

where $F_{000} = Z$, the total number of electrons in a unit cell. Equation (1.52) cannot be solved as it is, since $F_{obs}(k)$ defined on $\rho(r)$. In order to solve equation (1.52) in a simple manner, we introduce the following approximation which replaces $F_{cal}(k)$ as

$$F_{cal}(H) = V \sum \tau(r) \, exp\left(-2\pi i \vec{k}.\vec{r}\right) dV \tag{1.53}$$

This approximation is called zeroth order single pixel approximation (ZSPA). By using this approximation, the right hand side of equation (1.52) becomes independent of τ(r) and it can be solved in an iterative way starting from a given initial density for the prior distribution. As the initial density for the prior density τ(r), a uniform density distribution is employed in this work,

$$\tau(r) \leq \tau(r) \geq \frac{Z}{M} \tag{1.54}$$

where M is the number of pixels for which the electron density is calculated. The reason for this choice of prior distribution is that the uniform density distribution corresponds to the maximum entropy state among all possible density distributions. In the calculation of $\rho(r)$, all of the symmetry recruitments are satisfied and the number of electrons (Z) is always kept constant through an iteration process. Mathematically, the summation concerning $\rho(r)$ in the above equations should be written as an integral. Since we must use a very limited number of pixels in the numerical calculation, the integral is replaced by the summation in the above equations.

The major difficulties of traditional Fourier synthesis methods of extracting electron densities from powder data are,

- Termination due to unobserved peaks, which produce spurious features, making it difficult to extract physically meaningful residual electron density distribution maps

- The difficulty in obtaining accurate observed structure factors (F_{obs}), owing to the collapse of reciprocal space onto a one dimensional diffraction pattern and the resulting partial or complete overlap of Bragg reflections

MEM method has overcome these two difficulties. Due to its reproducibility of the true charge density distribution in a unit cell, this method has enormous respect from the crystallographers. It is possible to evaluate the reflections missing from the summation after completion of the MEM enhancement. The amplitudes of the unobserved reflections are assumed to be equal to zero in a Fourier summation, while the MEM technique (Collins, 1982) provides the most probable values. The maximum entropy method (MEM) can give accurate structure analysis than any other methods.

1.8.2 Advantages of maximum entropy method

The advantages of maximum entropy method (MEM) over the conventional Fourier method are,

- It is an appropriate method which requires limited amount of information by maximizing information entropy under constraints to handle the uncertainties properly

- It acquires electron density distribution which is consistent with observed physical quantities

- This method yields least biased information and hence MEM electron densities are always positive even with limited number of data sets

- The calculated electron density distribution provides the detailed structure information without using structural model

- The bonding nature and the distribution of electrons in the bonding region can be clearly visualized using this technique

- It can give accurate structure analysis when compared to other conventional methods

1.8.3 Computational procedure of MEM

The computational procedure for the determination of charge density was proposed by Takata et al. (2001). Figure 1.10 shows the virtual imaging lens by means of a computer program as in MEM.

The MEM program read an input file with the crystal parameters, reflection list with real and imaginary parts of the structure factor, an asymmetric unit density file and a documentation file. The MEM (Collins, 1982) refinements were carried out by dividing the unit cell into suitable pixels. The uniform prior density was used in all the cases by dividing the total number of electrons by the volume of the unit cell. This model provides information on reconstructing the structure factor using preliminary information like position, type, space group, etc. The calculated structure factor is then compared with the observed one and the resultant calculated structure factor and observed one will be used for the reconstruction of charge density using MEM analysis (Collins, 1982). These charge density pictures are compared at each stage in iterative cycles based on satisfying the constrained C. The error is then determined and then a revised model of the charge density is again projected. This model will replace the previous model and then iteratively, the structure factor can be evaluated. The Lagrange parameter is suitably chosen so that the convergence criterion C=1 is reached after minimum number of iterations. At this stage, the charge density is analyzed for its bonding and charge ordering behavior.

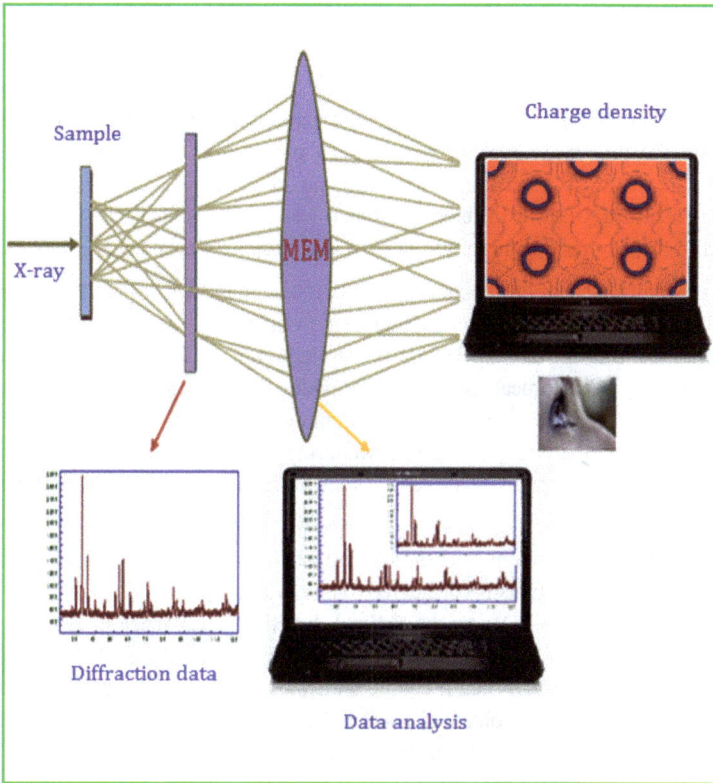

Figure 1.10 MEM virtual imaging lens by means of a computer program.

In the present work, the software package PRIMA (Izumi and Dilanian, 2002) has been used for the numerical MEM computations and the program VESTA (Momma and Izumi, 2006) has been used for the pictorial visualization of the three dimensional (3D) and two dimensional (2D) representation of the electron densities.

1.9   Local structure analysis by atomic pair distribution function (PDF)

1.9.1   Atomic pair distribution function (PDF)

The approach of traditional crystallographic structure solution is no longer sufficient to understand the properties of complex materials on an atomic scale. The limitation of traditional structure refinements yields only the long range average structure of the material. In recent years, a different approach has been developed that can tackle the structural issues in many of these systems to a high degree of accuracy.

The atomic pair distribution function (PDF) is a technique which has greatly improved our ability to find local structural deviations from a well defined average structure (Egami, 1990; Billinge et al., 1996). In crystallographic methods, Bragg-peaks are analyzed directly in reciprocal space which provides extensive and sufficient information about the underlying structure. The PDF technique and closely related total scattering methods allow both the Bragg and diffuse scattering to be analyzed together without bias, revealing the short and intermediate range order of the material, regardless of the degree of disorder.

Fourier relationship between measurable diffraction intensities and the real-space arrangement of pairs of atoms is considered to be advantage of the PDF technique (Egami and Billinge, 2003). The PDF can be defined directly in real-space in terms of atomic coordinates. The probability of finding an atom at a given radial distance r from another atom is provided by the PDF i.e., it can be understood as a bond length distribution. On the other hand, the PDF is obtained by the use of the Fourier transform of the total scattering pattern and yields the local structure on length scales beyond the unit cell size.

It is possible to get high-quality data at high-Q values, allowing accurate high real-space resolution PDFs to be determined (Petkov et al., 1999), if we use high energy X-rays. It was thought that neutrons were superior for high-Q measurements because, as a result of the Q dependence of the X-ray atomic form factor, the X-ray coherent intensity gets rather weak at high Q; however, the high flux of X-rays from modern synchrotron sources more than compensates for this and we have shown that high quality high-resolution PDFs can be obtained using X-rays.

A PDF spectrum (Bragg peaks + diffuse scattering) consists of a series of peaks, the positions of which give the distances of atom pairs in real space. The relative thermal atomic motion and static disorder is responsible for the ideal width of these peaks. The investigation of effects of lattice vibration on PDF peak width is important for the following two reasons.

- To establish the degree to which information on phonons and the inter-atomic potentials can be obtained from powder diffraction data
- To account for correlation effects in order to extract information on static disorder in a disordered system such as an alloy

1.9.2 Practical aspects of pair distribution function

The PDF method illustrates that it agrees well with the inter-atomic distances computed from a crystallographic model (average model) (Toby and Egami, 1992; Peterson et al., 2003), when there are no short range deviations from the average structure, This real space method is one of the experimental techniques that can be used to probe structures on the nanometer length scale, when the local structure is not consistent with the long range, globally averaged structure (Egami and Billinge, 2003). The scattered X-ray intensity by a collection of atom after corrections for absorption, polarization, multiple scattering and normalization to the unit of one atom or scattering s can be expressed as,

$$I(Q) = \sum_{ij} f_i(Q) f_j(Q) \langle \exp[iQ.(r_i - r_j)] \rangle \tag{1.55}$$

where $f_i(Q)$ is the scattering amplitude of single atom i, r_i is the position of the ith atom, $\langle \exp[iQ.(r_i - r_j)] \rangle$ is the quantum and thermal average $\vec{Q}$ is given by,

$$\vec{Q} = k_f - k_i \tag{1.56}$$

$$\vec{Q} = |\vec{Q}| = 2|k_{f,i}| sin\theta \qquad (if \ |k_f| = |k_i|) \tag{1.57}$$

where k_f and k_i are the momenta of the scattered and incident photons or neutrons, respectively and θ is the diffraction angle. Inelastic phonons scattering is also included in the quasi-elastic X-ray scattering. The average structure factor can be given as,

$$S(\vec{Q}) = \frac{I(\vec{Q})}{<f(\vec{Q})>^2} + \frac{[<f(\vec{Q})>^2 - <f(\vec{Q})^2>]}{<f(\vec{Q})>^2} \tag{1.58}$$

Here $< f(\vec{Q}) >^2$ is the compositional average. This equation for $S(\vec{Q})$ can be noted as the square of the structure factor $F(\vec{Q})$.

The total data will be treated in real space, by Fourier transforming the data, instead of indexing and analyzing each powder peak separately. The atomic pair distribution function (PDF) can be obtained from the Fourier transform of equation (1.58)

$$\rho(r) = \rho_0 + \frac{1}{2\pi^2 r} \vec{Q}[S(\vec{Q}) - 1] \sin(\vec{Q}r) \, d\vec{Q} \tag{1.59}$$

where ρ_0 is the average (atomic) number density (average no of atoms in the sample with respect to distance) and r is the distance.

If $\vec{Q}$ is in $\mathring{A}^{-1}$, then $\rho(r)$ will be in $\mathring{A}^{-3}$. $\rho(r)$ corresponds to the (atomic) number density at a distance r from the average atom (Warren, 1990). Equation (1.59) is expressed as;

$$\rho(r) - \rho_0 = \frac{1}{2\pi^2 r} \vec{Q}[S(\vec{Q}) - 1] \sin(\vec{Q}r) \, d\vec{Q} \tag{1.60}$$

$$4\pi r[\rho(r) - \rho_0] = \frac{2}{\pi} \vec{Q}[S(\vec{Q}) - 1] \sin(\vec{Q}r) \, d\vec{Q} \tag{1.61}$$

The $G(r)$ is an experimentally accessible function. It is referenced with respect to the average density ρ_0. The experimental PDF is a direct Fourier transform of the total scattering structure function $S(\vec{Q})$, the corrected, normalized intensity from powder scattering data $I(Q)$, given by

$$G(r) = 4\pi r[\rho(r) - \rho_0] = \frac{2}{\pi} \int_{Q=0}^{Q=Q_{max}} \vec{Q}[S(\vec{Q}) - 1] \sin(\vec{Q}r) \, dQ \tag{1.62}$$

$G(r)$ will be in units of $\mathring{A}^{-2}$ if $\vec{Q}$ is the magnitude of the wave vector in $\mathring{A}^{-1}$. The function $G(r)$ gives information about the number of atoms in a spherical shell of unit thickness at a distance r from a reference atom. It peaks at characteristic distances separating pairs of atoms and thus reflects the atomic structure.

The structure function is related to the coherent part of the total diffracted intensity of the

$$S(\vec{Q}) = 1 + \frac{I^{coh}(\vec{Q}) - \sum c_i |f_i(\vec{Q})|^2}{|\sum c_i f_i(\vec{Q})|^2} \tag{1.63}$$

where $I^{coh}(Q)$ is the coherent scattering intensity per atom in electron units and c_i and f_i are the atomic concentration and X-ray scattering factor, respectively, for the atomic species of type i (Warren, 1990; Egami and Billinge, 2003). $G(r)$ is simply another representation of the diffraction data. However, exploring the diffraction data in real space has advantages especially in the case of materials with significant structural disorder. In order to refine an experimental PDF, one needs to calculate a PDF from a structural model. This can be done using the relation

$$G_{calc}(r) = \frac{1}{r} \sum_i \sum_j \left[\frac{b_i b_j}{\langle b \rangle^2} \delta(r - r_{ij}) \right] - 4\pi r \rho_0 \qquad (1.64)$$

where the sum goes over all pairs of atoms i and j within the model separated by r_{ij}. The scattering power of atom i is b_i and $\langle b \rangle$ is the average scattering power of the sample. b_i is the scattering length in neutron scattering, in case of X-rays, it is the atomic form factor evaluated at a user defined value of Q. The Bragg and diffuse scattering information about the local atomic arrangements in a material can be obtained through the atomic pair distribution function. The experimental PDF was obtained as follows;

- First, the coherently scattered intensities were extracted from X-ray diffraction pattern by applying appropriated correction for flux, background, Compton scattering and sample absorption

- The intensities were normalized in absolute electron unit, reduced to structure function and Fourier transformed to atomic PDF

- The powder X-ray intensities collected were normalized to obtain the total scattering function, using the software PDFgetX (Jeong, 2001)

- The experimental PDF peak widths as a function of pair distance are extracted using the "real space" Rietveld program PDFGUI (Farrow et al., 2007), which allows for powerful usability features such as real time plotting and remote execution of the fitting program whilst visualizing the results locally

The structural parameters like, lattice parameters, phase scale factor, linear atomic correlation factor, quadratic atomic correlation factor, spherical nano-particle amplitude correction, low r peak sharpening, peak sharpening cut off and cut off for profile setup functioning were refined to get the absolute phase in the refinement program PDFfit (Proffen and Billinge, 1999). The step size, data scale factor, upper limit for Fourier transform to obtain data PDF, resolution damping factor, resolution peak broadening factor, data collection temperature and doping concentration levels etc., are the data configuration parameters to be refined to get accurate PDF fitting. Finally, the observed and calculated PDF's are visualized and compared.

1.10 A review of the NLO materials chosen for the present investigation

1.10.1 Lead molybdate (PbMoO$_4$)

PbMoO$_4$ has been the subject of extensive study because of its potential applications, such as photoconductivity (Bernhardt and Jena, 2006), luminescence (Hizhnyi and Nedilko, 2003), thermo-luminescence (Bollmann, 2006) and photo catalysis (Kudo et al., 1990). The optical properties of lead molybdate have been reported by Tyagi et al. (2008). The influence of lattice defects to improve the optical and other properties of PbMoO$_4$ crystals have been reported by Piwowarska et al. (2008). But, a study on electron density distribution of PbMoO$_4$ is not available in the literature. Therefore, an attempt has been made in this research work to study the electron density distribution using the X-ray powder diffraction data.

1.10.2 Lithium niobate (LiNbO$_3$)

Lithium niobate (LiNbO$_3$) is one of the most investigated NLO materials for widespread and promising applications in non-linear optics (e.g., parametric amplification, second-harmonic generation, holographic data storage, and optical information processing) (Arizmendi, 2004). LiNbO$_3$ can be utilized as a high temperature acoustic transducer, such as an accelerometer for jet aircraft because of its high Curie temperature (T$_c$ =1210° C) (Radonjic et al., 2005). The existence of threshold effect with regard to magnesium doping level in LiNbO$_3$ single crystal has been confirmed by measuring IR absorption and photoconductivity by Sweeney et al. (1984). It was reported that the high-speed and low-noise holographic storage with considerate diffraction efficiency has been achieved in LiNbO$_3$ crystal by using two dopants Mg and Fe (i.e., Mg:Fe:LiNbO$_3$) (Xu et al., 2005). LiNbO$_3$ crystals doped with magnesium might lead to the formation of defects and suppress the optical damage (Feng et al., 1990). The holographic storage properties of the LiNbO$_3$ crystal can be greatly improved by doping with ZnO (Volk et al., 1990). In spite of the fact that the single crystals of LiNbO$_3$ have several applications, still there are restrictions because of their high cost and difficult fabrication. Kar et al. (2004) have reported the growth of LiNbO$_3$ using Czochralski technique. In the present work, an attempt has been made to study the electron density distribution, the bonding between the atoms and local structure of LiNbO$_3$.

1.10.3 Ce doped gadolinium gallium garnet (Ce:Gd$_3$Ga$_5$O$_{12}$)

The early work on non-linear optical materials showed that gadolinium gallium garnet (Gd$_3$Ga$_5$O$_{12}$, GGG), is a subject of intense research for the past three decades, because of its interesting properties, such us chemical stability, mechanical hardness, good thermal and optical behavior (Yu et al., 2008). It was reported that the GGG single crystals can be used as an appropriate substrate for yttrium iron garnet (YIG) and YIG-like magneto-optical epitaxial film in the field of integrated optics (Shiraishi, 1985). Neodymium doped GGG single crystal (Nd:GGG) can be regarded as a promising and key material for solid-state high power heat capacity laser (Albrecht et al., 1998). A method to increase the lattice parameters of GGG crystal employing coupled substitution of gallium by magnesium and zirconium, and gadolinium by calcium has been reported (Sugimoto et al., 1996). A study on electron density distribution of $Ce:Gd_3Ga_5O_{12}$ is not available in the literature. Hence, this material is chosen for the electron density analysis in this work.

1.10.4 Calcite (CaCO₃)

Calcite ($CaCO_3$) is one of the most probed NLO materials for widespread and promising applications. It was reported that calcite crystal has strong uniaxial anisotropic nature (Thompson et al., 1998). Numerous studies are available on the pressure induced phase transition behavior of $CaCO_3$. It was reported that $CaCO_3$ undergo phase transitions from calcite I to calcite II (slightly denser phase), and from calcite II to calcite III (significantly denser phase), at a pressure of 1.44 GPa and 1.77 GPa respectively (Bridgman, 1939). It was reported that the above mentioned phase transitions occur at 1.45 GPa and 1.74 GPa respectively at room temperature (Singh and Kennedy, 1974). It was also reported that the same phase transitions occur at 1.5 GPa and 2.2 GPa respectively (Merrill and Bassett, 1975). A comprehensive non-destructive methodology for the simultaneous quantitative determination of the calcite crystal phases using the Fourier transform Raman spectroscopy (FT-RS) was reported by Christos and Nikos (2000). The present work on calcite ($CaCO_3$) can be considered as a clear and precise attempt in visualizing the electron density distribution and bonding nature in the unit cell.

1.10.5 Yb doped calcium fluoride (Yb:CaF₂)

Calcium fluoride (CaF_2) crystal has been reported as a good non-linear optical material because of its excellent properties, such as, broad transmittance range (from far UV to mid-IR), low refractive index, high chemical resistance and high laser damage threshold (Li et al., 2000). It was reported that the fluoride crystals were preferred when compared to oxides, because of their lower refractive index limiting non-linear effects and low phonon energy reducing non-radiative relaxation (Liangbi et al., 2005). CaF_2 crystal was reported as one of the first host materials for possible lasing action (Baker, 1974). The synthesis and upconversion luminescence properties of $Er^{3+}:CaF_2$ nanoparticles co-doped with Yb^{3+} was reported (Pedroni et al., 2011). It was reported that the Yb doped CaF_2 crystal has less thermal dispersion coefficient (dn/dT) than Yb doped YAG crystal, suitable for high power fundamental-mode laser operation (Boudeile et al., 2008). Rare

Earth doped CaF_2 crystals have proved their usefulness in photolithography (Jain et al., 1982) and medical and biological labeling (Yang et al., 2010). Though Yb doped CaF_2 material was reported as a good non-linear optical material, the electron level properties were not available in the literature. Hence, this material is chosen for the electron density analysis in this work.

1.10.6 Aluminium oxide (Al_2O_3), $Cr:Al_2O_3$ and $V:Al_2O_3$

Aluminium oxide (Al_2O_3) is one of the most analysed non-linear optical materials. Its hardness, excellent dielectric properties, refractoriness and good thermal properties make it the material of choice, for a wide range of applications (Zhao et al., 2004). It was reported that the refractive index of Al_2O_3 is suitable for optical waveguide (Kersten et al., 1975). It was also reported that Al_2O_3 crystal undergoes a variety of transitions until the most stable α structure (α-Al_2O_3) was formed at a temperature above 1000°C (Bahlawane et al., 2000). The present work on Al_2O_3, $Cr:Al_2O_3$ and $V:Al_2O_3$ can be considered as an attempt in visualizing the electron density distribution and the local structure.

1.11 Scope of the present work

From the review of the literature, it is found that the available non-linear optical materials have widespread and promising applications, but the efforts to understand the local structure, electron density distribution and bonding is still lacking.

The present research work has been carried out to explore the structural details, the electron density distribution and the local bond length distribution of some non-linear optical materials only (not as the characterization of non-linear properties of the optical materials). The present study has also been concerned with the estimation of the optical band gap, the particle size, crystallite size, and the elemental composition from UV-Visible analysis, SEM, XRD and EDS of some non-linear optical materials respectively. The materials chosen for the present analysis are:

- Lead molybdate ($PbMoO_4$)
- Lithium niobate ($LiNbO_3$)
- Ce doped Gadolinium gallium garnet ($Ce:Gd_3Ga_5O_{12}$)
- Calcite ($CaCO_3$)
- Yb doped Calcium fluoride ($Yb:CaF_2$)
- Al_2O_3, $Cr:Al_2O_3$ and $V:Al_2O_3$

In the present work, the experimental X-ray powder diffraction data sets of the chosen non-linear optical materials have been subjected to the Rietveld (1969) refinement technique with the help of the software JANA2006 (Petříček et al., 2006). The refined structure factors have been used to explicate the fine details of the electron density distribution inside the unit cell using the maximum entropy method (MEM) with the help of the software PRIMA (Izumi and Dilanian, 2002). The three-dimensional, two-dimensional electron density maps and one-dimensional electron density profiles have been constructed and analyzed using the software program VESTA (Visualization for Electronic and STructural Analysis) (Momma and Izumi, 2006). The bond length distributions of the selected non-linear optical materials have been estimated using PDF (Petkov et al., 1999). The atomic pair distribution function (PDF) is obtained from the Fourier transform of the measured X-ray powder diffraction data using the software PDFgetX (Jeong et al., 2001). The observed and calculated PDF have been compared using the graphical software PDFgui (Farrow et al., 2007). The particle sizes and crystallite size of the chosen non-linear optical materials have been evaluated using SEM and XRD respectively. The elemental composition analyses of the chosen non-linear optical materials have been performed by using EDS. The optical band gaps of the chosen non-linear optical materials have also been estimated by using UV-Visible spectroscopy.

References

[1]   Albrecht G. F., Sutton S. B., George E. V., Sooy W. R and Krupke W. F., Laser Part Beams, Vol. 16, pp. 605, 1998. https://doi.org/10.1017/S0263034600011435

[2]   Arizmendi L., Phys. Status Solidi (a), Vol. 201, pp. 253, 2004. https://doi.org/10.1002/pssa.200303911

[3]   Atteberry J, 'How Scanning Electron Microscopes Work', April 2014, Online:HowStuffWorks.com.<http://science.howstuffworks.com/scanning-electron microscope.htm> 06 April 2014

[4]   Bahlawane N and Watanabe T., Journal of the American Ceramic Society, Vol. 83(9), pp. 2324, 2000. https://doi.org/10.1111/j.1151-2916.2000.tb01556.x

[5]   Baker J. M., Crystals with the fluorite structure, Oxford press, Clarendon, 1974

[6]   Bernhardt H and Jena R. S., Phys. Status Solidi (a), Vol. 64, pp. 207, 2006. https://doi.org/10.1002/pssa.2210640122

[7]   Billinge S. J. L., DiFrancesco R. G., Kwei G. H., Neumeier J. J and Thompson J. D., Phys. Rev. Lett., Vol. 77, pp. 715, 1996. https://doi.org/10.1103/PhysRevLett.77.715

[8]   Bollmann W, Krist. Tech., Vol. 15, pp. 367, 2006. https://doi.org/10.1002/crat.19800150320

[9]   Boudeile J., Didierjean J., Camy P., Doualan J. L., Benayad A., Menard V., Moncorge R., Druon F., Balembois F and Georges P., Opt. Express, Vol. 16(14), pp. 10098, 2008. https://doi.org/10.1364/OE.16.010098

[10]   Bridgman P. W., Am. J. Sci., Vol. 237, pp. 7, 1939. https://doi.org/10.2475/ajs.237.1.7

[11]   Caglioti G., Paeletti A and Ricci F. P., Nucl. Instrum., Vol. 3, pp. 223, 1958. https://doi.org/10.1016/0369-643X(58)90029-X

[12]   Cheetham A. K and Taylor J. C., J. Solid State Chem., Vol. 21, pp. 253, 1977. https://doi.org/10.1016/0022-4596(77)90203-1

[13]   Christos G. K and Nikos V. V., Analyst, Vol. 125, pp. 251, 2000. https://doi.org/10.1039/a908609i

[14]   Collins D. M., Nature, Vol. 298, pp. 49, 1982. https://doi.org/10.1038/298049a0

[15]   Cox D. E., Hastings J. B., Thomlinson W and Prewill C. T., Nucl. Instrum. Method, Vol. 208, pp. 573, 1983. https://doi.org/10.1016/0167-5087(83)91185-7

[16]   Cullity B. D and Stock S. R., Elements of x-ray diffraction, Third edition, chapter 14, Prentice Hall, 2001

[17]   Dollase W. A., J. Appl. Cryst., Vol. 19, pp. 267, 1986. https://doi.org/10.1107/S0021889886089458

[18]   Eckhardt G., Hellwarth R. W., McClung F. J., Schwarz S. E., Weiner D and Woodbury E. J., Phys. Rev. Lett., Vol. 9, pp. 455, 1962. https://doi.org/10.1103/PhysRevLett.9.455

[19]   Egami T., Mater. Trans., JIM, Vol. 31, pp.163, 1990

[20]   Egami T and Billinge S. J. L., Underneath the Bragg Peaks: Structural Analysis of Complex Material, Oxford University Press, London, 2003

[21]   Farrow C. L., Juhas P., Liu J. W., Bryndin D., Bozin E. S., Bloch J., Proffen Th and Billinge S. J. L., J. Phys., Condens. Matter, Vol.19, pp. 335219, 2007. https://doi.org/10.1088/0953-8984/19/33/335219

[22] Feng X., Wang D and Zhang J., Phys. Status Solidi(b), Vol. 157, pp. 127, 1990. https://doi.org/10.1002/pssb.2221570239

[23] Franken P. A., Hill A. E, Peters C. W and Weinreich G., Phys. Rev. Lett., Vol. 7(4), pp. 118, 1961. https://doi.org/10.1103/PhysRevLett.7.118

[24] Gilmore C. J., Acta Cryst., Vol. A52, PP. 561, 1996

[25] Giordmaine J. A and Miller R. C., Phys. Rev. Lett., Vol. 14, pp. 973, 1965. https://doi.org/10.1103/PhysRevLett.14.973

[26] Gull S. F and Daniel G. J., Nature, Vol. 272, pp. 686, 1978. https://doi.org/10.1038/272686a0

[27] Gullapalli S and Barron A. R, 'Characterization of Group 12-16 (II-VI) Semiconductor Nanoparticles by UV-visible Spectroscopy', OpenStax CNX, June, 2010; Online: Web site. http://cnx.org/content/m34601/1.1/

[28] Hewat A. W., Chem. Scripta, Vol. 26A, pp. 119, 1986

[29] Howard C. J., Journal of Applied Crystallography, Vol. 15, pp. 615, 1982. https://doi.org/10.1107/S0021889882012783

[30] Hizhnyi Y. A and Nedilko S. G., J. Lumin., Vol. 102, pp. 688, 2003. https://doi.org/10.1016/S0022-2313(02)00625-7

[31] International Tables for X-ray Crystallography, The Kynoch press, Birmingam, England, 1974

[32] Itoh T., Shirasaki S., Fujie Y., Kitamura N., Idemoto Y., Osaka K., Ofuchi H., Hirayama S., Honma T and Hirosawa I., Journal of Alloys and Compounds, Vol. 491, pp. 527, 2010. https://doi.org/10.1016/j.jallcom.2009.10.262

[33] Izumi F and Dilanian R.A., Recent Research Developments in Physics, Vol. 3, Part II, Transworld Research Network, Trivandrum, pp.699, 2002

[34] Jain K., Willson C. G and Lin B. J, IEEE electron device letters, Vol. 3, pp. 3, 1982. https://doi.org/10.1109/EDL.1982.25476

[35] Jaynes E. T., IEEE Transactions on Systems Science and Cybernetics, Vol. SSC-4, pp. 227, 1968. https://doi.org/10.1109/TSSC.1968.300117

[36] Jeong K., Thompson J., Proffen Th., Perez A and Billinge S. J. L., PDFGetX, A Program for Obtaining the Atomic Pair distribution function from X-ray Powder Diffraction Data, 2001

[37]   Kar S., Bhatt R., Bartwal K. S and Wadhawan V. K., Cryst. Res. Technol., Vol. 39, pp. 230, 2004. https://doi.org/10.1002/crat.200310175

[38]   Kersten R. T., Mahlein H. F and Rauscher W., Thin Solid Films, Vol. 369, pp. 28, 1975

[39]   Kudo A., Steinberg M., Bard A. J., Campion A., Fox M. A., Mallouk T. E., Webber S. E and White J. M., Catal. Lett., Vol. 5, pp. 61, 1990. https://doi.org/10.1007/BF00772094

[40]   Larson A. C and Von Dreele R. B, General Structure Analysis System, Los Alamos National Laboratory, Los Alamos, 1990

[41]   Liangbi Su, Jun Xu, Hongjun, Lei Wen, Yueqin Zhu, Zhiwei Zhao, Yongjun Dong, Guoqing Zhou, Jiliang Si, Chemical Physics Letters, Vol. 406, 254, 2005

[42]   Li Y. H and Jiang G. J., J. Synth. Cryst., Vol. 29, pp. 221, 2000

[43]   Malmros G and Thomas J. O., J. Appl. Crystallogr., Vol. 10, pp. 7, 1977. https://doi.org/10.1107/S0021889877012680

[44]   Momma K and Izumi F, Journal of Applied Crystallography, Vol. 44, pp. 1272, 2011. https://doi.org/10.1107/S0021889811038970

[45]   Merrill L and Bassett W. A., Acta Cryst. B, Vol.31, pp. 343, 1975. https://doi.org/10.1107/S0567740875002774

[46]   Patterson A., Phys. Rev., Vol. 56, pp. 978, 1939. https://doi.org/10.1103/PhysRev.56.978

[47]   Pedroni M., Piccinelli F., Passuello T., Giarola M., Polizzi S., Bettinelli M and Speghini A., Nanoscale, Vol. 3, pp. 1456, 2011. https://doi.org/10.1039/c0nr00860e

[48]   Peterson P. F., Bozin E. S, Proffen Th and Billinge S. J. L., J. Appl. Cryst., Vol. 36, pp.53, 2003. https://doi.org/10.1107/S0021889802018708

[49]   Petkov V., Jeong I. K., Chung J. S., Thorpe M. F., Kycia S and Billinge S. J. L., Phys. Rev. Lett., Vol. 83, pp.4089, 1999. https://doi.org/10.1103/PhysRevLett.83.4089

[50]   Petříček V., Dušek M and Palatinus L., JANA 2006, The crystallographic computing system, Institute of Physics, Academy of sciences of the Czech republic, Praha, 2006

[51]   Piwowarska D., Kaczmarek S. M and Berkowski M., J. Non-Cryst. Solids, Vol. 354, pp. 4437, 2008. https://doi.org/10.1016/j.jnoncrysol.2008.06.078

[52] Prince E, The Rietveld Method, Ed. by Young R. A, Oxford University Press, 1993

[53] Proffen Th. and Billinge S. J. L., J. Appl. Cryst., Vol. 32, pp. 572, 1999. https://doi.org/10.1107/S0021889899003532

[54] Radonjic L., Todorovic M and Miladinovic J., Mater. Sci. Eng. B, Vol. 121, pp. 64, 2005. https://doi.org/10.1016/j.mseb.2005.03.003

[55] Rietveld H. M., J. Appl. Crystallogr., Vol. 2, pp. 65, 1969. https://doi.org/10.1107/S0021889869006558

[56] Saravanan R., Private communication

[57] Scherrer P, Nachr. Ges. Wiss. Göttingen., Vol. 26, pp. 98, 1918 (in German)

[58] Schrodinger E., Proc. Royal Irish Academy, Vol. 47 A, pp. 77, 1942

[59] Schrodinger E., Proc. Royal Irish Academy, Vol. 49 A, pp. 59, 1943

[60] Shen Y. R., The Principles of Nonlinear Optics, Wiley, New York, 1984

[61] Shiraishi K., Appl. Opt., Vol. 24, pp. 951, 1985. https://doi.org/10.1364/AO.24.000951

[62] Singh A. K and Kennedy G. C., J. Geophys. Res., Vol. 79, pp. 2615, 1974. https://doi.org/10.1029/JB079i017p02615

[63] Sugimoto N., Terui H., Tate A., Katoh Y., Yamada Y., Sugita A., Shibukawa A and Inoue Y., J. Lightwave Technol., Vol. 14, pp. 2537, 1996. https://doi.org/10.1109/50.548152

[64] Sweeney K. L., Halliburton L. E., Bryan D. A., Rice R. R., Gerson R and Tomaschke H. E., Appl. Phys. Lett., Vol. 45, pp. 805, 1984. https://doi.org/10.1063/1.95372

[65] Takata M., Nishibori E and Sakata M., Z. Kristallogr., Vol. 216, pp. 71, 2001

[66] Terhune R. W., Maker P. D and Savage C. M., Phys. Rev. Lett., Vol. 8, pp. 404, 1962. https://doi.org/10.1103/PhysRevLett.8.404

[67] Thompson D. W., Devries M. J., Tiwald T. E and Woollam J. A., Thin Solid Films, Vol. 313, pp. 341, 1998. https://doi.org/10.1016/S0040-6090(97)00843-2

[68] Thompson P, Cox D. E and Hastings J. B, Journal Applied Crystallography, Vol. 20, pp. 79, 1987. https://doi.org/10.1107/S0021889887087090

[69] Toby B. H and Egami T., Acta Cryst., Vol. A48, pp. 336, 1992. https://doi.org/10.1107/S0108767391011327

[70]   Tyagi M., Sangeeta C. T., Desai D. G and Sabharwal S. C., J. Lumin., Vol. 128, pp. 22, 2008. https://doi.org/10.1016/j.jlumin.2007.05.005

[71]   Volk T. R., Pryalkin V. I and Rubinina N. M., Opt. Lett., Vol. 15, pp. 996, 1990. https://doi.org/10.1364/OL.15.000996

[72]   Warren B. E., X-ray Diffraction, chapter 3, Dover publications, New York, 1990

[73]   Xu Z., Xu S., Xu Y and Wang R., Proc. SPIE Int. Soc. Opt. Eng., Vol. 5636, pp. 505, 2005

[74]   Yang J., Deng Y., Wu Q., Zhou J., Bao H., Li Q., Zhang F., Li F., Tu B and Zha D., J. Langmuir, Vol. 26, 8850, 2010

[75]   Young R. A., The Rietveld method, Oxford University Press, 1993

[76]   Young R. A., Mackie, P. E and Von Dreele R. B., J. Appl. Crystallogr., Vol. 10, pp. 262, 1977. https://doi.org/10.1107/S0021889877013466

[77]   Young R. A and Desai P., Archiwum Nauki of Materialach, Vol. 10 (1-2), pp. 71, 1989

[78]   Yu F. P., Yuan D. R., Duan X. L., Guo S. Y., Wang X. Q., Cheng X. F and Kong L. M, J. Alloys Compd., Vol. 465, pp. 567, 2008. https://doi.org/10.1016/j.jallcom.2007.11.008

[79]   Zhao Z. W., Tay B. K., Yu G. Q., Chua D. H. C., Lau S. P and Cheah L. K., Thin Solid Films, Vol. 14, pp. 447, 2004

[80]   Zernike F and Midwinter J., Applied Nonlinear Optics, Wiley, New York, 1973

Chapter 2

Powder X-ray Analysis of Non-Linear Optical Materials

Abstract

Chapter II provides the methods of sample preparation and the analysis of the observed powder X-ray diffraction data sets collected using these samples of non-linear optical materials such as, $PbMoO_4$, $LiNbO_3$, $Ce:Gd_3Ga_5O_{12}$, $CaCO_3$, $Yb:CaF_2$, and Al_2O_3, $Cr:Al_2O_3,V:Al_2O_3$. The results of fitting PXRD profiles for all the non-linear optical materials are reported and analyzed.

Keywords

PXRD, Rietveld Refinement, $PbMoO_4$, $LiNbO_3$, $Ce:Gd_3Ga_5O_{12}$, $CaCO_3$, $Yb:CaF_2$, Al_2O_3, $Cr:Al_2O_3,V:Al_2O_3$

Contents

2.1   Introduction

In the field of materials science, materials are generally classified into two types as crystalline and amorphous materials. In crystalline materials, the atoms are arranged in a periodic pattern with equal interplaner/interatomic distance and which leads to have a specified structure. The wavelength of X-rays used to probe the crystal structure is almost comparable with the interatomic distance in crystalline materials. Hence, X-rays can only be used in the electromagnetic spectrum to study the structure of the crystalline materials. Hence, X-ray characterization technique is most suitable for analyzing the structural information in the crystalline materials. The intensities obtained from the diffracted monochromatic X-ray beams at a particular angle give the structural and phase information for the chosen system. X-ray diffraction technique can also be used to identify the unknown structures from powder materials. In the present work, powder X-ray diffraction technique has been used for determining the structure of the chosen non-linear optical materials.

Rietveld (1967) refinement technique based on the least square refinement method has been adopted for analysing the structural and phase information for the observed X-ray data set. Rietveld (1967) refinement technique is an effective technique for determining the detailed structural information of the crystalline materials and the results obtained from this technique can be used for constructing the charge density distribution in the unit cell. In order to analyze the structural information for the chosen system, JANA 2006 software (Petříček et al., 2006) which uses Rietveld (1967) method was utilized in the present research work.

The present work has been carried out on non-linear optical materials to elucidate the properties using electron density distribution studies through maximum entropy method (Collins, 1982). A detailed structural analysis of the materials is required to analyze the electron density distribution. This analyzes the derived structural information of the non-linear optical materials. The non-linear optical materials chosen for the present analysis are Lead molybdate ($PbMoO_4$), Lithium niobate ($LiNbO_3$), Ce doped $Gd_3Ga_5O_{12}$ ($Ce:Gd_3Ga_5O_{12}$), Calcite ($CaCO_3$), Yb doped Calcium fluoride ($Yb:CaF_2$) and Al_2O_3, Cr doped Al_2O_3 ($Cr:Al_2O_3$) and V doped Al_2O_3 ($V:Al_2O_3$).

2.2 Lead molybdate ($PbMoO_4$)

2.2.1 Powder X-ray data analysis of $PbMoO_4$

High purity analytical grade (Alfa Aesar, 99.9%) $PbMoO_4$, powder purchased from the scientific suppliers has been used for the present analysis. The powder X-ray intensity data set of $PbMoO_4$ were collected in the 2θ range from $10°$ to $120°$ with step size $0.017°$ at the National Institute for Interdisciplinary Science and Technology (NIIST), CSIR, Trivandrum, India, using an X'-PERT PRO (Philips, Netherlands), X-ray diffractometer with a monochromatic incident beam of wavelength 1.54056 Å, offering pure CuK_α radiation. The raw X-ray intensity data of $PbMoO_4$ was refined using Rietveld (1967) refinement, a well known powder profile fitting method for the structural refinement. The cell parameters and other structural parameters refined using this method, by JANA2006 software (Petříček et al., 2006) are given in table 2.2.1 and table 2.2.2. The coordinate of Pb, Mo and O atoms are (0, 0, 0.5), (0, 0, 0) and (0.3749, 0.0133, 0.0726) respectively. Refined structural parameters of $PbMoO_4$ are given in this table 2.2.1 and the deviations from the reported values (Gull and Daniel, 1978) are reasonable. The Rietveld (1967) refined powder profile of $PbMoO_4$ using JANA 2006 (Petříček et al., 2006) is shown in figure 2.2.1.

Table 2.2.1 Refined structural parameters of $PbMoO_4$.

Parameter	Value
a (Å)	5.4503(28)
b (Å)	5.4503(28)
c (Å)	12.1431(63)
R_p (%)	13.30
$_wR_p$ (%)	19.17
R_{obs} (%)	7.14
$_wR_{obs}$ (%)	6.62
GOF	0.41

R_P - Reliability index for profile

$_wR_P$ - Weighted reliability index for profile

R_{obs} - Reliability index for observed structure factors

$_wR_{obs}$ - Weighted reliability index for observed structure factors

GOF - Goodness of Fit

49

Table 2.2.2 The observed and calculated structure factors of $PbMoO_4$.

h	k	l	F_{obs}	F_{cal}	$\sigma(F_{obs})$
1	0	1	98.4573	98.7006	2.8429
1	1	2	349.7430	349.9710	1.3312
1	0	3	87.7628	87.2345	1.3324
0	0	4	309.7490	310.2990	1.5862
2	0	0	462.5380	461.1600	0.7041
2	0	2	41.1272	41.6542	1.1429
2	-1	1	89.6529	88.4611	1.7989
2	1	1	109.9920	108.7153	1.6607
1	1	4	75.1904	75.2219	1.1016
1	0	5	131.5640	133.0430	3.7179
2	-1	3	111.3520	111.2460	2.3168
2	0	4	337.8600	339.3370	3.0845
2	2	0	344.9590	340.9680	3.4480
3	0	1	86.6224	85.7185	1.6367
3	-1	2	367.6080	369.1780	2.9506
2	0	6	92.6194	95.0129	1.4878
0	0	8	493.949	494.6710	1.8810
4	0	0	582.8250	581.8400	1.1826
4	-1	1	15.7574	15.6825	0.3942
4	-1	3	126.4850	131.0940	2.9979
4	2	0	303.0800	304.8420	1.4298
3	3	4	33.5254	32.3107	1.7131
1	1	10	297.6380	294.2270	3.2072
3	1	8	90.9099	90.0081	1.1528
4	-3	1	70.8994	67.2338	2.7338
5	1	2	307.6050	308.8360	1.3492
0	0	12	377.5650	374.3680	2.6534
3	1	10	238.5420	235.3290	1.9762
4	-2	8	260.3300	265.9020	1.3651
5	-3	2	316.2010	322.7780	1.4985
4	0	10	20.2954	20.8786	2.3240

F_{obs} - observed structure factor

F_{cal} - calculated structure factor

$\sigma(F_{obs})$ - standard error in the measurements

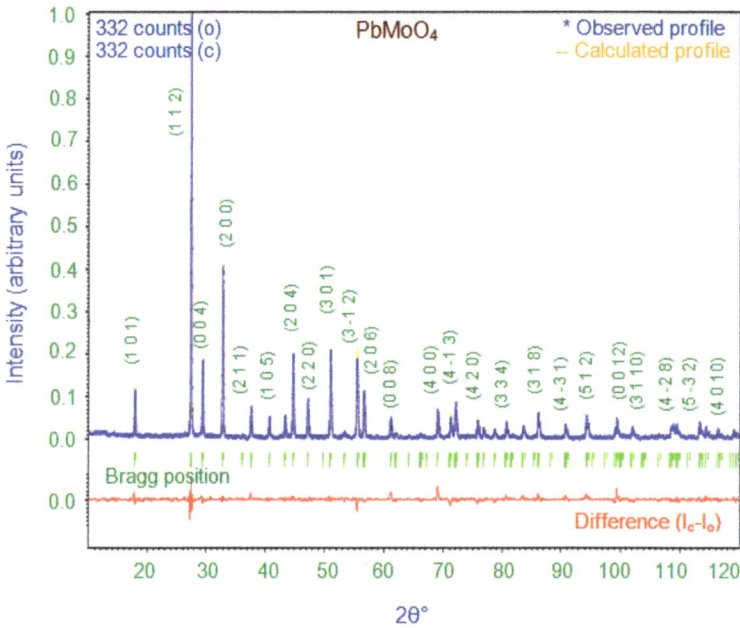

Figure 2.2.1 Refined powder XRD profile of PbMoO₄.

2.3 Lithium niobate (LiNbO₃)

2.3.1 Sample preparation of LiNbO₃

Lithium niobate (LiNbO$_3$) crystal has been grown with the technical expertise and advice from the Department of Applied Physics, Graduate School of Engineering, Tohoku University, Sendai, Japan, by Czochralski method. LiNbO$_3$ crystal has been grown from the melt with a composition 51.4/48.6 for the molar ratio of Nb$_2$O$_5$/ Li$_2$CO$_3$, which is the congruent melt composition. Starting materials of Nb$_2$O$_5$ and Li$_2$CO$_3$ with high purities were weighed according to the composition at the congruent melting point, mixed and kept in a furnace for solid state reaction at 1000°C for 24 hours. This process has been repeated to ensure the completion of the solid-state reaction. Then the product has been ground, and made into pellets. The growth has been accomplished using a procedure similar to Kar et al (2004).

2.3.2 Powder X-ray analysis of LiNbO₃

A small portion of grown LiNbO$_3$ crystal was crushed properly for powder XRD study. The powder X-ray intensity data set was collected in order to analyze the bonding and structural behaviour of LiNbO$_3$ in the 2θ range from 10° to 120° with step size 0.1° at Regional Research Laboratory (RRL), CSIR, Trivandrum, India, using an X'-PERT PRO (Philips, Netherlands), X-ray diffractometer with a monochromatic incident beam of wavelength 1.54056 Å, offering pure CuK$_\alpha$ radiation. The raw intensity data set of LiNbO$_3$ was refined using Rietveld (1967) refinement. It is a well known powder profile fitting method for the structural refinement. The cell parameters and other structural parameters were refined by using the software program JANA 2006 (Petříček et al., 2006). The refined structural parameters of LiNbO$_3$ using JANA 2006 (Petříček et al., 2006) are given in table 2.3.1. The refined structure factors of LiNbO$_3$ are tabulated in table 2.3.2. The powder XRD data fitted using JANA 2006 (Petříček et al., 2006) is shown in figure 2.3.1.

Table 2.3.1 Refined structural parameters of LiNbO₃.

Parameter	Value
a (Å)	5.1588(13)
b (Å)	5.1588(13)
c (Å)	13.8910(39)
R_p (%)	10.89
$_wR_p$ (%)	15.01
R_{obs} (%)	3.39
wR_{obs} (%)	3.28
GOF	1.11

R_P - Reliability index for profile

$_wR_P$ - Weighted reliability index for profile

R_{obs} - Reliability index for observed structure factors

$_wR_{obs}$ - Weighted reliability index for observed structure factors

GOF- Goodness of Fit

Table 2.3.2 The observed and calculated structure factors of LiNbO₃.

h	k	l	F_{obs}	F_{cal}	$\sigma(F_{obs})$
1	-1	2	229.3390	230.4240	2.0056
1	0	4	143.3240	145.5700	2.7524
2	-1	0	129.2520	126.8270	3.0863
0	0	6	77.0896	74.2805	4.8109
2	-1	3	68.0691	68.4391	2.5749
2	0	2	104.7560	110.5580	3.7796
2	-2	4	212.8960	212.1630	4.0785
2	-1	6	152.4990	150.6660	3.2196
3	-1	1	23.2662	24.9324	2.2841
1	-1	8	104.6130	105.6280	5.1283
3	-1	4	120.4260	116.3870	4.0772
3	0	0	167.7020	157.5170	6.0684
3	-2	5	22.2635	31.0188	3.8703
2	0	8	125.4280	124.4890	6.8797
1	0	10	91.5134	98.2450	6.2894
2	-1	9	29.5202	31.6896	2.0283
4	-2	0	94.7524	90.7140	7.5236
3	-1	7	34.2471	32.7874	2.7193
3	0	6	90.1434	94.6964	4.2523
3	-3	6	123.8870	130.1440	5.844
4	-2	3	25.2081	26.1405	1.3835
4	-1	2	115.0820	117.5980	4.8316
3	-2	8	119.6450	121.8810	4.3518
4	-3	4	105.0160	104.7710	6.2497
0	0	12	35.20340	41.32660	3.9812
4	-1	5	31.89120	34.5160	1.6298
4	-2	6	125.9860	131.7630	5.4406

F_{obs} - observed structure factor

F_{cal} - calculated structure factor

$\sigma(F_{obs})$ - standard error in the measurements

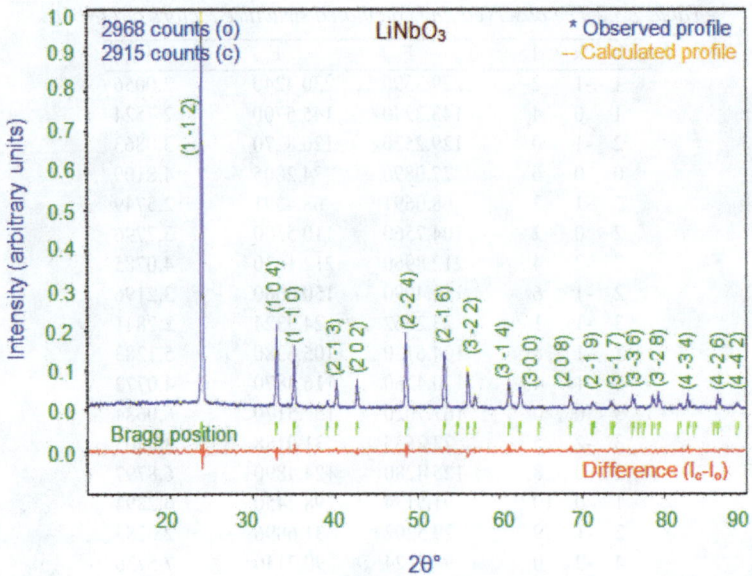

Figure 2.3.1 Refined powder XRD profile of LiNbO₃.

2.4 Gadolinium gallium garnet (Gd₃Ga₅O₁₂)

2.4.1 Sample preparation of Gd₃₋ₓCeₓGa₅O₁₂

Cerium doped gadolinium gallium garnet ($Gd_{3-x}Ce_xGa_5O_{12}$) powder has been grown with the technical proficiency and guidance from the Department of General and Inorganic Chemistry, Vilnius University, Naugarduko, Vilnius, Lithuania by an aqueous sol-gel method. The molar percentages of Ce used in Ce:GGG was 0.5, 1 and 3. In the aqueous sol-gel process, the following materials were used: Gd_2O_3 (99.99%, Aldrich), Ga_2O_3 (99.9 %, Merck) and $(NH_4)_2Ce(NO_3)_6$ (99.9 %, Merck). Gd and Ga nitrate solutions were prepared by dissolving the corresponding oxides in 65% nitric acid at 60-65°C. The obtained nitrates were washed several times with distilled water. The appropriate quantities of $[NH_4]_2[Ce(NO_3)_6]$ were dissolved separately in 100 ml of distilled water. Clear solutions were obtained after stirring at 60°C-65°C for 1 hour in beakers covered with a watch-glass. The resulting mixtures were stirred at 65°C for 1 hour, followed by dropwise addition of tris-(hydroxymethyl) aminomethane ($C_4H_{11}NO_3$) as complexing agent. The resulting solutions were mixed at the same temperature for 1 hour and then concentrated by slow solvent evaporation at 65°C until they turned into transparent gels.

The gels were dried in an oven at 110°C for 24 hours. The resulting gel powders were ground in an agate mortar and heated in air at 800°C for 5 hours by slow temperature elevation (~3-4°C/minute). After grinding in an agate mortar, the powders were further sintered in air at 1000 °C for 10 hours.

2.4.2 Powder X-ray analysis of $Gd_{3-x}Ce_xGa_5O_{12}$

Powder X-ray intensity data sets were collected for $Gd_{3-x}Ce_xGa_5O_{12}$ (x = 0.5, 1 and 3) materials in 2θ range from 10° to 120° with step size 0.02° at Sophisticated Analytical Instruments Facility (SAIF), Department of Science and Technology (DST), Cochin, India, using a Bruker AXS D8 Advance (Karlsruhe, Germany). Pure CuK_α radiation with a wavelength of 1.54056 Å was used as the incident beam. The observed X-ray data sets for $Gd_{3-x}Ce_xGa_5O_{12}$ (x = 0.5, 1 and 3) are shown in figure 2.4.1. The enlarged view of (204) peak shows a shifting of diffraction angle (2θ) towards lower angle with respect to Ce concentration. This angular shifting leads to increase in the cell parameters. The increasing trend in cell parameters is shown in table 2.4.1. The raw intensity data as seen from figure 2.4.1 is in decreasing trend with Ce concentration due to lower atomic number of dopant atoms (Ce) than that of the host atom (Gd). This reveals that the dopant atom has been successfully incorporated into the lattice as substitutional defects. The raw intensity data of $Gd_{3-x}Ce_xGa_5O_{12}$ (x = 0.5, 1 and 3) were refined using Rietveld (1967) refinement technique which was proposed by Rietveld (1967). The Rietveld (1967) method is an exact tool to find out many structural details. The structural parameters such as lattice parameters, peak shift, back ground parameters, profile shape parameters, preferred orientation parameters and etc are refined and are extracted from this method. In this work, the X-ray diffraction data of cubic $Gd_{3-x}Ce_xGa_5O_{12}$ (x = 0.5, 1 and 3) material is refined using JANA2006 software (Petříček et al., 2006) by considering the space group of $Ia\bar{3}d$. The refined structural parameters from the powder data sets using JANA2006 (Petříček et al., 2006) of $Gd_{3-x}Ce_xGa_5O_{12}$ (x = 0.5, 1 and 3) are summarized in table 2.4.1. The refined structure factors of $Gd_{3-x}Ce_xGa_5O_{12}$ (x = 0.5, 1 and 3) are tabulated in tables 2.4.2(a) to 2.4.2(c) respectively. A standard software package formulated by Holland and Redfern (1997) was also used to refine the cell parameters of $Gd_{3-x}Ce_xGa_5O_{12}$ (x = 0.5, 1 and 3) using the observed 2θ values. The refined cell parameter increases with Ce (r_i=1.034 Å) (Shannon, 1976) concentration due to large ionic radius than that of the host atom Gd (r_i=0.938 Å) (Shannon, 1976). It infers that inclusion of cerium content certainly increases the lattice parameters of gadolinium gallium garnet which plays a significant role in the optical properties of non-linear optical material. The fitted XRD profiles along with their difference are shown in figures 2.4.2(a) to 2.4.2(c). The Rietveld (1967) refinement fitting again confirms the phase pure system of gadolinium gallium garnet.

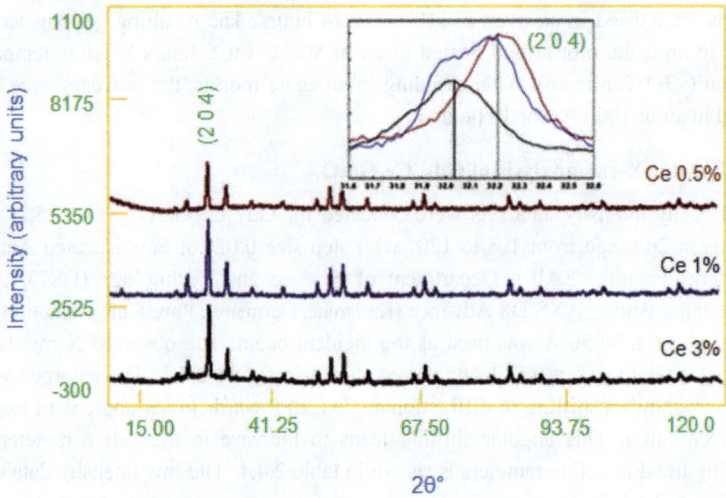

Figure 2.4.1 Observed X-ray profiles of $Gd_{3-x}Ce_xGa_5O_{12}$ (x = 0.5, 1 and 3).

Table 2.4.1 Refined structural parameters of $Ce:Gd_3Ga_5O_{12}$.

Refined parameter	Composition of Ce		
	0.5	1	3
a=b=c (Å)	12.413(25)	12.434(14)	12.460(21)
α=β=γ	90°	90°	90°
V (Å³)	1912.96	1922.74	1934.58
R_p (%)	3.52	3.76	3.60
$_wR_p$ (%)	4.48	4.79	4.56
R_{obs} (%)	4.97	6.21	3.99
$_wR_{obs}$ (%)	4.79	4.52	3.64
GOF	1.09	1.11	1.03

V - Volume of the unit cell

R_P - Reliability index for profile

$_wR_P$ - Weighted reliability index for profile

R_{obs} - Reliability index for observed structure factors

$_wR_{obs}$ - Weighted reliability index for observed structure factors

GOF - Goodness of Fit

Table 2.4.2(a) *The observed and calculated structure factors of $Gd_{3-x}Ce_xGa_5O_{12}$ for x =* 0.5.

h	k	l	F_{obs}	F_{cal}	$\sigma(F_{obs})$
1	1	2	166.9230	181.8140	4.0380
2	1	3	168.1930	178.1190	4.6446
0	0	4	938.5180	928.3350	6.4772
2	0	4	1074.8200	1066.0800	9.4127
2	2	4	789.6060	779.7100	7.3527
2	1	5	258.2190	252.7400	6.9608
1	1	6	289.8890	291.3320	6.8819
4	4	4	1062.1400	1065.4800	7.5133
4	0	6	957.8820	954.7260	5.2181
4	2	6	759.2500	754.1780	6.3150
0	0	8	1366.8400	1348.3200	6.5758
4	0	8	637.8050	632.0900	4.6434
4	2	8	724.6180	721.1670	4.0949
6	1	7	101.1310	108.0270	3.9893
5	5	6	152.5300	164.1870	6.5201
2	1	9	124.8710	130.8830	3.8937
6	4	6	591.2990	591.0670	3.0799
6	5	7	194.7260	191.2630	8.5131
4	0	10	596.4610	593.6230	6.3726
6	1	9	126.6600	133.3550	2.7752
3	3	10	136.2510	143.2630	2.8697
4	2	10	507.7850	506.0090	2.2928
8	0	8	1019.2900	1017.6000	4.8695
8	1	9	22.0247	22.2541	1.8207
4	3	11	0.4391	0.4328	0.0294
7	4	9	10.4014	10.5097	0.8598
6	4	10	421.6200	422.5300	2.7044
3	1	12	13.7984	14.4266	0.6206
8	3	9	4.3039	4.5064	0.2106
5	3	12	17.3419	18.1425	0.4738
8	4	10	491.9770	491.8570	4.8651
4	3	13	3.7903	4.0032	0.3452
7	1	12	3.0213	3.1910	0.2752
8	7	9	4.3475	4.7310	0.4348
8	3	11	2.9678	3.1792	0.2831

F_{obs} - observed structure factor

F_{cal} - calculated structure factor

$\sigma(F_{obs})$ - standard error in the measurements

Table 2.4.2(b) The observed and calculated structure factors of $Gd_{3-x}Ce_xGa_5O_{12}$ for $x = 1$.

h	k	l	F_{obs}	F_{cal}	$\sigma(F_{obs})$
1	1	2	165.8430	185.0490	6.0954
2	1	3	183.0790	189.4090	7.1300
0	0	4	894.2060	901.0040	8.6276
2	0	4	1106.5000	1099.0500	9.2360
2	2	4	760.8980	770.2290	6.0637
2	1	5	264.5200	254.2060	7.5509
1	1	6	309.8030	310.0080	6.4267
4	4	4	1074.1800	1055.9500	5.8038
4	0	6	950.7990	948.6750	3.9419
4	2	6	733.2730	745.1080	8.5513
0	0	8	1426.8600	1498.2400	7.4033
4	0	8	666.0360	676.1700	6.4163
4	2	8	729.5970	736.5630	4.0875
6	1	7	104.2950	101.1250	5.8753
5	5	6	138.7220	136.9640	6.4874
2	1	9	113.4420	111.0290	1.5343
6	4	6	578.6540	574.0300	6.8859
6	5	7	161.5000	163.0650	6.3663
4	0	10	682.4960	687.4300	4.2743
6	1	9	138.1670	138.2290	3.0316
3	3	10	108.3780	109.3980	4.2625
4	2	10	543.6990	545.4970	5.0266
8	0	8	1172.6400	1078.0200	5.6193
8	1	9	29.5222	30.3334	5.5823
4	3	11	16.7629	19.3854	6.0501
7	4	9	22.7253	23.3497	4.2971
6	4	10	477.467	471.2990	4.7355
3	1	12	3.0936	3.0283	0.3902
8	3	9	4.4032	4.2951	0.2646
5	3	12	15.9578	18.5295	1.4156
8	4	10	571.653	517.9040	3.2738
4	3	13	12.0662	12.2253	1.7131
7	1	12	10.5027	12.0076	1.3590
8	7	9	5.7143	6.2343	0.5064
8	3	11	9.5174	10.8589	1.0437

F_{obs} - observed structure factor

F_{cal} - calculated structure factor

$\sigma(F_{obs})$ - standard error in the measurements

Table 2.4.2(c) *The observed and calculated structure factors of* $Gd_{3-x}Ce_xGa_5O_{12}$ *for* $x = 3$.

h	k	l	F_{obs}	F_{cal}	$\sigma(F_{obs})$
1	1	2	209.0510	203.4530	2.8315
2	1	3	160.4000	174.3830	9.0511
0	0	4	941.9790	943.7390	6.9395
2	0	4	1055.2300	1060.1000	9.0407
2	2	4	759.5400	764.4990	8.4352
2	1	5	278.5230	260.4030	7.2203
1	1	6	219.9650	215.7000	7.2519
4	4	4	1089.9500	1076.2500	6.7434
4	0	6	972.9710	973.5220	4.3054
4	2	6	761.5220	756.8390	8.9926
0	0	8	1464.7200	1487.7900	8.4281
4	0	8	656.9360	660.9030	6.8426
4	2	8	684.1200	687.7150	5.9325
6	1	7	99.3252	100.3910	5.5565
5	5	6	112.4430	118.4700	3.3840
2	1	9	159.1060	168.4960	4.4338
6	4	6	607.3480	606.8070	2.6104
6	5	7	112.2610	115.8170	6.7226
4	0	10	609.0730	609.4850	1.0517
6	1	9	137.0090	136.6050	4.7103
3	3	10	123.3500	125.4050	2.2146
4	2	10	537.1320	536.8630	4.5006
8	0	8	1041.9500	1041.8000	2.2146
8	1	9	5.8533	5.4923	0.2555
4	3	11	46.6551	44.1114	2.0219
7	4	9	7.6449	7.2281	0.3313
6	4	10	416.1390	415.632	1.2324
3	1	12	31.4055	33.9459	0.8526
8	3	9	1.9909	2.1524	0.0540
5	3	12	29.5728	30.3551	0.7309
8	4	10	500.9510	500.576	1.9843
4	3	13	18.9440	19.9013	1.8974
7	1	12	10.1053	11.0091	1.3992
8	7	9	7.0048	7.4115	0.7136
8	3	11	3.6163	3.8380	0.8570

F_{obs} - observed structure factor

F_{cal} - calculated structure factor

$\sigma(F_{obs})$ - standard error in the measurements

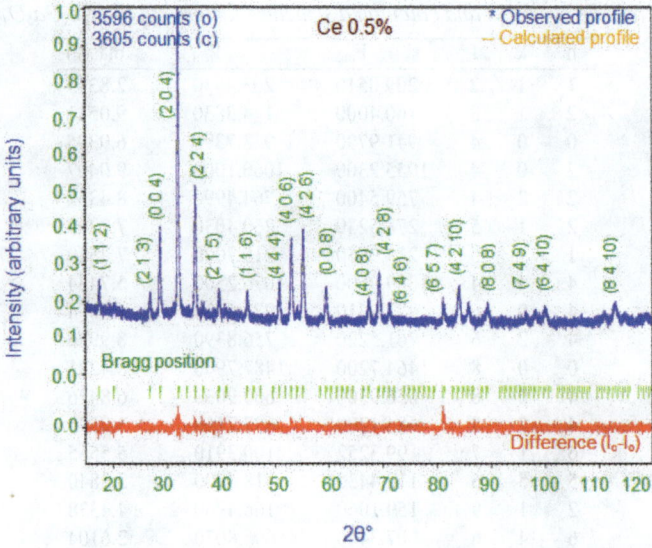

Figure 2.4.2(a) Refined powder XRD profile of $Gd_{3-x}Ce_xGa_5O_{12}$ for x = 0.5.

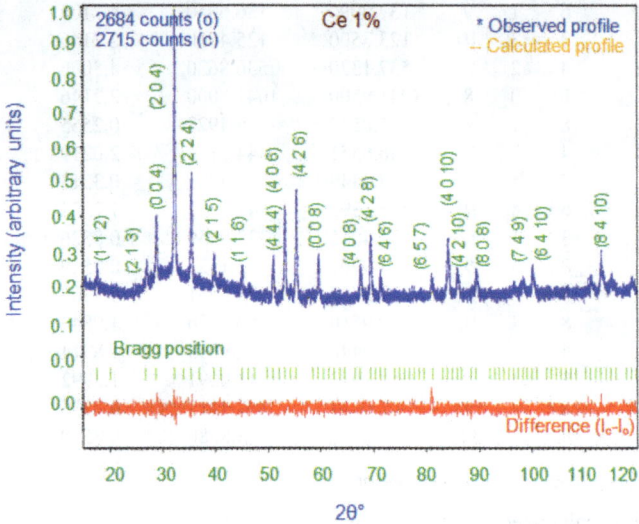

Figure 2.4.2(b) Refined powder XRD profile of $Gd_{3-x}Ce_xGa_5O_{12}$ for x = 1.

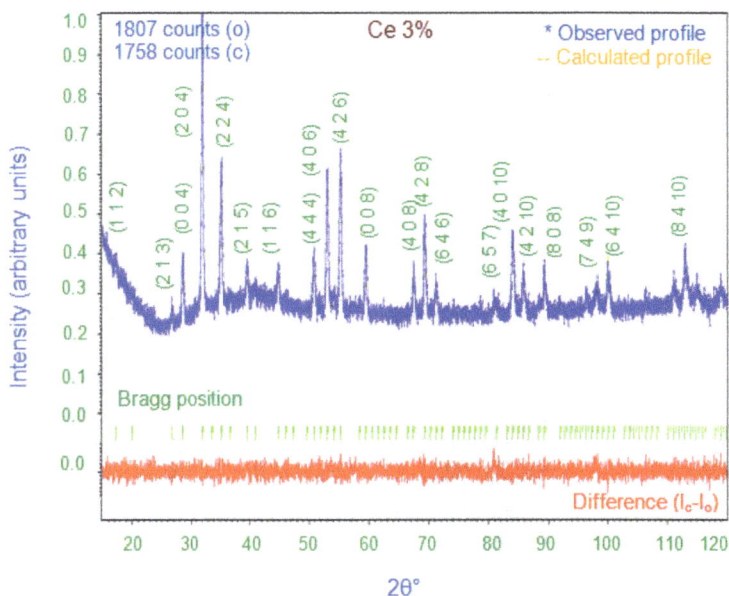

Figure 2.4.2(c) Refined powder XRD profile of $Gd_{3-x}Ce_xGa_5O_{12}$ for x = 3.

2.5 Calcite ($CaCO_3$)

2.5.1 Sample preparation of $CaCO_3$

Mechanical milling was applied for 5 hrs in the single crystal calcite material using an agate mortar. The prepared sample was characterized by X-ray diffractometer (XRD). The X-ray intensity data set was collected at Sophisticated Analytical Instruments Facility (SAIF), Department of Science and Technology (DST), Cochin, India, using a parallel-beam Bruker AXS D8 Advance (Karlsruhe, Germany), X-ray diffractometer fitted with Si (Li) detector. Soller silt set to a 6° (2θ) aperture was used to improve the peak shape and θ-2θ geometry. The accelerating voltage and the applied current density were 40 KV and 35 mA/cm^2 respectively. The wavelength used for the X-ray intensity data collection was 1.54056 Å, offering pure CuK$_\alpha$ radiation with a 2θ range of data collection from 5° to 120° with a step size of 0.01° and the counting time of 49.2s at each step.

2.5.2 Powder X-ray analysis of $CaCO_3$

The raw intensity data set was refined based on the Rietveld (1967) method using the software program JANA2006 (Petříček et al., 2006). The Rietveld (1967) method refines structural parameters like fractional co-ordinates, atomic displacement parameters, occupation factors and lattice parameters from the whole powder diffraction patterns. In this method, the observed profile is matched with the profile constructed using the pseudo-Voigt profile shape function of Thompson et al. (1987) which was modified to some extent to accommodate various Gaussian FWHM parameters and the Scherrer co-efficient P for Gaussian broadening. The asymmetric parameters are refined using Berar-Baldinozzi function employing the multi-beam Simpson rule integration devised by Howard (1982). A correction for preferred orientation of the crystallites in the sample is dealt with the model as proposed by March-Dollase (March, 1932; Dollase, 1986). The Legendre polynomial of first kind was used to fit the background. The fitted XRD profile along with their difference is shown in figure 2.5.1. The atomic positional parameters, cell parameters and structural parameters from Rietveld (1967) method with the corresponding standard deviations in parenthesis are given in tables 2.5.1 and 2.5.2. The observed and calculated structure factors of calcite are given in table 2.5.3.

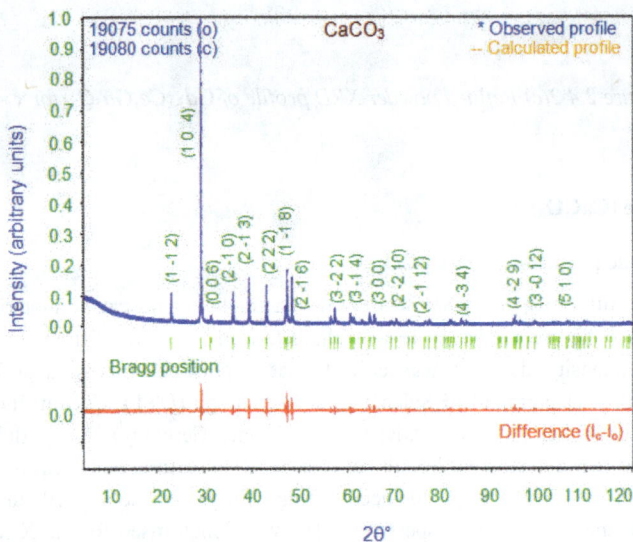

Figure 2.5.1 Refined powder XRD profile of $CaCO_3$.

Table 2.5.1 Atomic positions and thermal vibration parameters of CaCO₃.

Parameter	Ca	C	O
x	0	0	0.2534(2)
y	0	0	0
z	0	0.25(1)	0.25(2)

x, y, z - positional parameters

Table 2.5.2 Refined structural parameters of CaCO₃.

Parameter	Value
a (Å)	4.9912 (9)
b (Å)	4.9912 (9)
c (Å)	17.0717(11)
R_p (%)	7.47
$_wR_p$ (%)	11.19
R_{obs} (%)	3.56
$_wR_{obs}$ (%)	4.24
GOF	1.34

R_P - Reliability index for profile

$_wR_P$ - Weighted reliability index for profile

R_{obs} - Reliability index for observed structure factors

$_wR_{obs}$ - Weighted reliability index for observed structure factors

GOF- Goodness of Fit

Table 2.5.3 The observed and calculated structure factors of CaCO₃.

h	k	l	F_{obs}	F_{cal}	$\sigma(F_{obs})$
1	-1	2	32.3370	32.7401	0.3696
1	0	4	144.1790	143.641	1.4554
0	0	6	36.2231	36.8761	0.5941
2	-1	0	64.9383	66.2716	0.7082
2	-1	3	55.8510	56.5845	0.5980
2	0	2	82.6861	81.6986	0.8974
1	-1	8	93.9592	98.6237	0.9952
2	-1	6	72.2177	73.9635	0.7661
3	-2	2	55.7422	59.2401	0.6539
3	-1	4	44.8548	46.1706	0.5662
2	0	8	41.8429	43.2733	0.5468
3	0	0	69.6160	73.0747	0.9219
0	0	12	94.7257	96.5906	1.3848
2	-2	10	46.4739	48.0525	0.8855
3	-2	8	36.1768	39.5446	0.6405
3	0	6	20.7354	21.4004	0.7715
2	-1	12	33.2823	33.2312	0.6794
4	-2	3	1.7325	1.6571	0.9007
4	-3	1	2.2753	2.4945	0.4410
4	-1	2	16.9585	16.9889	0.4141
3	-1	10	36.8837	38.0108	0.6908
1	-1	14	20.4805	21.2378	0.5901
4	-3	4	49.1524	53.7520	0.7364
4	-2	9	2.3205	2.3303	0.0346
1	0	16	34.6567	38.4724	0.8286
2	-1	15	13.1669	14.5875	0.3185
3	-1	13	12.4987	13.3727	1.0341
3	-3	12	41.2111	42.0875	0.8119
3	0	12	38.0279	38.8366	0.7492
5	-2	1	15.2263	15.7628	0.5941
5	-3	2	25.0272	25.8850	0.8062
4	-3	10	11.5444	11.5139	0.3232
4	-1	11	1.8021	1.9350	0.0521
0	0	18	4.1543	4.4914	0.1173
5	1	0	33.6958	35.3056	0.7435

F_{obs} - observed structure factor

F_{cal} - calculated structure factor

$\sigma(F_{obs})$ - standard error in the measurements

2.6 Calcium fluoride (CaF$_2$)

2.6.1 Sample preparation of Yb doped CaF$_2$

Ca$_{1-x}$Yb$_x$F$_2$ (x = 0, 0.03, 0.06, 0.09 and 0.12) materials have been synthesized by co-precipitation method according to Lyberis et al. (2001). Commercially available high pure calcium nitrate (99.98%, Alfa Aesar) and ytterbium nitrate (99.999%, Aldrich) have been used for the preparation of the desired materials. Calcium nitrate and ytterbium nitrate were dissolved in de-ionized water and added drop wise to a stirring solution of hydrofluoric acid. The precipitate was repeatedly washed in distilled water. The obtained powder was annealed at 400°C under argon atmosphere for 4 h. These samples were pelletized and sintered again under argon atmosphere at 900°C for 1 h.

2.6.2 Powder X-ray analysis of Yb doped CaF$_2$

The X-ray powder diffraction (XRPD) measurements of undoped and Yb doped CaF$_2$ materials were carried out at Sophisticated Analytical Instruments Facility (SAIF), Department of Science and Technology (DST), Cochin, India, using an Bruker AXS D8 Advance (Karlsruhe, Germany), X-ray diffractometer using pure CuK$_\alpha$ radiation. The wavelength used for the X-ray intensity data collection was 1.54056 Å, offering pure CuK$_\alpha$ radiation. The accelerating voltage and the applied current density were 40 KV and 35 mA/cm^2 respectively. Measurements were recorded with a 2θ range from 10° to 120° with step size of 0.02°. The observed X-ray peaks for the prepared CaF$_2$ materials match well with those of joint committee on powder diffraction standards (JCPDS) XRD data sets reported in the file (JCPDS Card No # 35-0816) and it has a cubic fluorite structure. All the peaks are indexed and given in figure 2.6.1(a). An important evidence for the perfect doping effect is that there is a shift in the diffracting angle 2θ towards lower angle with the increase of Yb concentration and it is evident through the enlarged XRD profile of the (1 1 3) plane as shown in the figure 2.6.1(b). The raw X-ray profiles of undoped and Yb doped CaF$_2$ materials have been refined for observing the structural changes due to the inclusion of Yb using Rietveld (1967) refinement technique which is employed in the software JANA 2006 (Petříček et al., 2006). The refinement was done considering cubic structure for CaF$_2$ which is having space group of $Fm\overline{3}m$ with four calcium atoms surrounded by each fluoride atom (Gerward et al., 1992). The structure of CaF$_2$ is identified with cell parameters a=b=c=5.4629 Å and the initial atomic positional coordinates for Ca and F are assumed as (0, 0, 0) and (0.25, 0.25, 0.25) respectively. The refined profiles of Ca$_{1-x}$Yb$_x$F$_2$ (x = 0, 0.03, 0.06, 0.09, 0.12) are shown in figures 2.6.2(a)-2.6.2(e) respectively. They show the perfect profile fitting between calculated and observed X-ray profiles. The refined structural parameters are tabulated in table 2.6.1 and the refined structure factors of Ca$_{1-x}$Yb$_x$F$_2$ (x = 0, 0.03, 0.06, 0.09, 0.12) are tabulated in tables 2.6.2(a)-2.6.2(e) respectively. The incorporation of Yb atoms in the preferential site of Ca atoms can also be revealed through the increase in the values of the structure factors reported in tables 2.6.2(a)-2.6.2(e). The increasing dopant concentration of ytterbium leads to the increase in cell dimension in the unit cell of CaF_2 as shown in table 2.6.1. The variation in cell dimension of $Ca_{1-x}Yb_xF_2$ (x = 0, 0.03, 0.06, 0.09 and 0.12) may not be correlated with the ionic radius, because the ionic radius of the dopant ions and the host ion are almost equal [r_i (Yb^{3+}) = 0.99 Å (Shannon, 1976), r_i (Ca^{2+}) = 1.0 Å (Shannon, 1976)]. When Ca^{2+} is substituted by a trivalent RE ion, charge compensating F^- ions enter the fluorite structure in interstitial fluorite cubic sites and the electronic repulsion between F^- ions leads to a net increase of the lattice parameter. Due to this reason, there is a shift in diffraction angles towards lower angle and is shown in figure 2.6.1(b). The intensities of the observed X-ray peaks for undoped and doped CaF_2 materials increase monotonically with the inclusion of Yb^{3+}. Due to the higher atomic number of Yb (Z_{Yb}=70, Z_{Ca}=20), the host lattice site gets higher number of electrons when doping is induced and it leads to increase in the X-ray intensity. The structure factors extracted from Rietveld (1967) refinement were used in the software PRIMA (Izumi and Dilanian, 2002) for construction of electron density distribution in the unit cell using MEM (Collins, 1982).

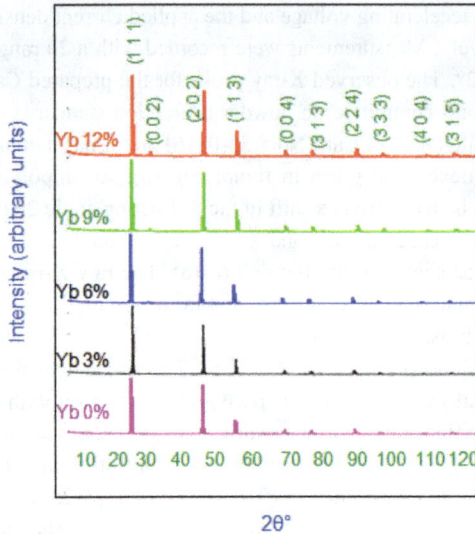

Figure 2.6.1(a) Observed XRD profiles of $Ca_{1-x}Yb_xF_2$ (x = 0.00, 0.03, 0.06, 0.09 and 0.12).

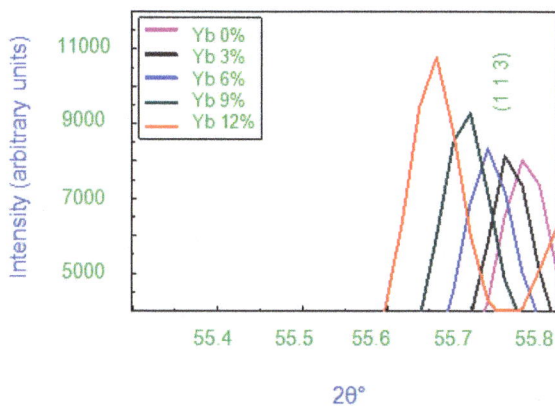

Figure 2.6.1(b) XRD profiles of (1 1 3) plane (enlarged) for Ca$_{1-x}$Yb$_x$F$_2$ (x = 0.00, 0.03, 0.06, 0.09 and 0.12).

Table 2.6.1 Refined structural parameters of Ca$_{1-x}$Yb$_x$F$_2$.

Parameters	x=0.00	x = 0.03	x = 0.06	x = 0.09	x = 0.12
a=b=c (Å)	5.4607(22)	5.465(4)	5.4681(6)	5.4710(7)	5.4743(12)
α=β= γ (°)	90	90	90	90	90
F$_{000}$	152	158	164	170	176
R$_p$ (%)	3.92	3.68	3.73	3.29	3.20
$_w$R$_p$ (%)	5.54	5.07	5.33	4.56	4.48
R$_{obs}$ (%)	0.98	1.35	1.13	1.21	1.47
$_w$R$_{obs}$ (%)	1.46	1.77	2.85	1.70	3.20
GOF	1.03	1.07	1.05	1.05	1.06

F$_{000}$ - Number of electrons in the unit cell

R$_P$ - Reliability index for profile

$_w$R$_P$ - Weighted reliability index for profile

R$_{obs}$ - Reliability index for observed structure factors

$_w$R$_{obs}$ - Weighted reliability index for observed structure factors

GOF - Goodness of Fit

Table 2.6.2(a) The observed and calculated structure factors of $Ca_{1-x}Yb_xF_2$ for $x = 0.00$.

h	k	l	F_{obs}	F_{cal}	$\sigma(F_{obs})$
1	1	1	61.9312	62.0552	0.6267
0	0	2	5.1399	6.1397	0.3102
2	0	2	91.7137	92.2070	0.9277
1	1	3	43.9111	44.1713	0.4568
0	0	4	66.3835	66.7216	0.7548
3	1	3	34.6845	34.7400	0.4016
2	2	4	51.4341	52.2789	0.5637
3	3	3	28.9634	29.0314	0.3639
4	0	4	42.7972	43.1778	0.5901
3	1	5	25.0015	25.1811	0.3186

F_{obs} - observed structure factor

F_{cal} - calculated structure factor

$\sigma(F_{obs})$ - standard error in the measurements

Table 2.6.2(b) The observed and calculated structure factors of $Ca_{1-x}Yb_xF_2$ for $x = 0.03$.

h	k	l	F_{obs}	F_{cal}	$\sigma(F_{obs})$
1	1	1	63.8081	66.9323	0.6478
0	0	2	8.7654	9.0040	0.3236
2	0	2	95.1839	96.0927	0.9677
1	1	3	47.9922	48.5645	0.5077
0	0	4	70.3746	71.3005	0.8703
3	1	3	38.2703	38.7483	0.4740
2	2	4	55.5760	56.5760	0.6354
3	3	3	32.6428	32.7433	0.4499
4	0	4	47.6879	47.2421	0.7712
3	1	5	28.4417	28.6629	0.4101

F_{obs} - observed structure factor

F_{cal} - calculated structure factor

$\sigma(F_{obs})$ - standard error in the measurements

Table 2.6.2(c) *The observed and calculated structure factors of* $Ca_{1-x}Yb_xF_2$ *for* x = 0.06.

h	k	l	F_{obs}	F_{cal}	$\sigma(F_{obs})$
1	1	1	70.6993	71.0204	0.7160
0	0	2	12.8850	13.8258	0.2792
2	0	2	100.1790	99.4771	1.0147
1	1	3	50.5921	50.8565	0.5256
0	0	4	70.8703	72.1547	0.8134
3	1	3	37.9468	39.8738	0.4328
2	2	4	53.0123	56.4015	0.5725
3	3	3	31.5784	33.0160	0.3865
4	0	4	44.3224	46.3040	0.5898
3	1	5	26.9385	28.2724	0.3255

F_{obs} - observed structure factor

F_{cal} - calculated structure factor

$\sigma(F_{obs})$ - standard error in the measurements

Table 2.6.2(d) *The observed and calculated structure factors of* $Ca_{1-x}Yb_xF_2$ *for* x = 0.09.

h	k	l	F_{obs}	F_{cal}	$\sigma(F_{obs})$
1	1	1	76.0336	75.7892	0.7700
0	0	2	18.9543	18.0037	0.3128
2	0	2	104.1820	104.3460	1.0554
1	1	3	55.2593	54.9629	0.5729
0	0	4	74.5199	76.6668	0.8519
3	1	3	43.2949	43.4845	0.4942
2	2	4	59.5191	60.5720	0.6504
3	3	3	36.6075	36.2458	0.4515
4	0	4	49.0087	50.1798	0.7073
3	1	5	31.3085	31.2002	0.3936

F_{obs} - observed structure factor

F_{cal} - calculated structure factor

$\sigma(F_{obs})$ - standard error in the measurements

Table 2.6.2(e) The observed and calculated structure factors of $Ca_{1-x}Yb_xF_2$ for x = 0.12.

h	k	l	F_{obs}	F_{cal}	$\sigma(F_{obs})$
1	1	1	80.2378	80.4675	0.8122
0	0	2	25.0889	23.0540	0.3390
2	0	2	106.7630	107.5370	1.0833
1	1	3	58.5778	58.8520	0.6089
0	0	4	75.6674	78.8955	0.8873
3	1	3	46.2350	46.8136	0.5319
2	2	4	61.5102	62.2076	0.6765
3	3	3	39.1166	39.1531	0.4747
4	0	4	51.5646	51.4039	0.7347
3	1	5	33.5296	33.7755	0.4128

F_{obs} - observed structure factor

F_{cal} - calculated structure factor

$\sigma(F_{obs})$ - standard error in the measurements

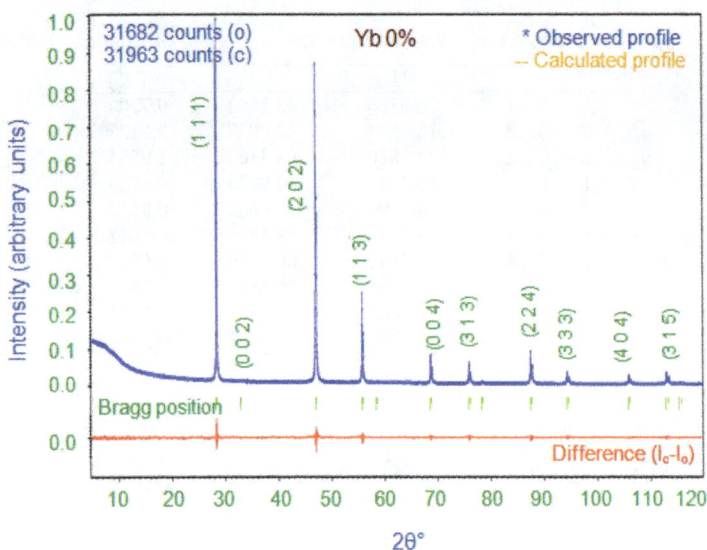

Figure 2.6.2(a) Refined powder XRD profile of $Ca_{1-x}Yb_xF_2$ for x = 0.00.

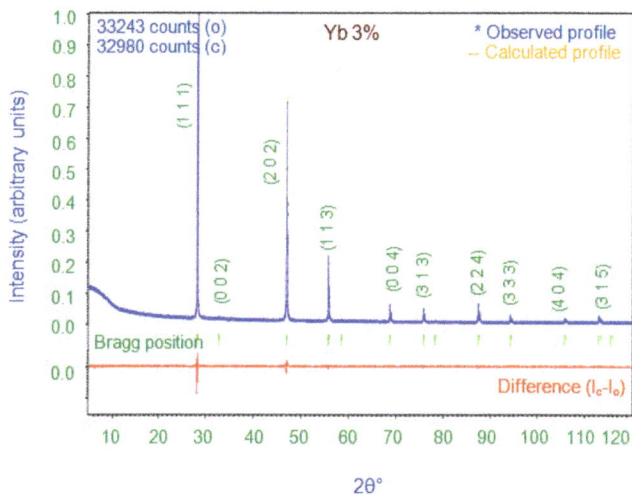

Figure 2.6.2(b) Refined powder XRD profile of $Ca_{1-x}Yb_xF_2$ for x = 0.03.

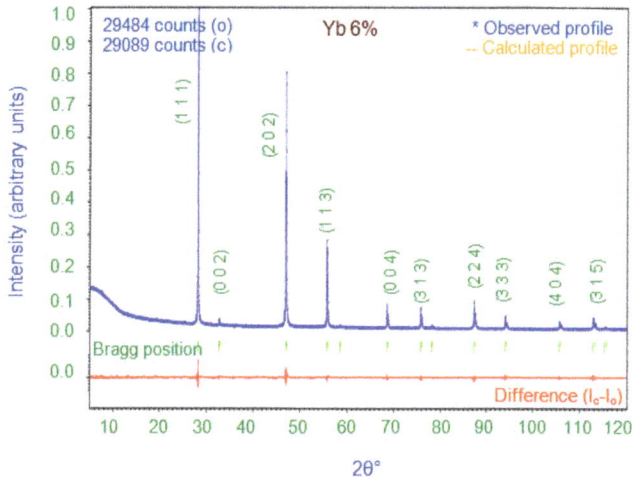

Figure 2.6.2(c) Refined powder XRD profile of $Ca_{1-x}Yb_xF_2$ for x = 0.06.

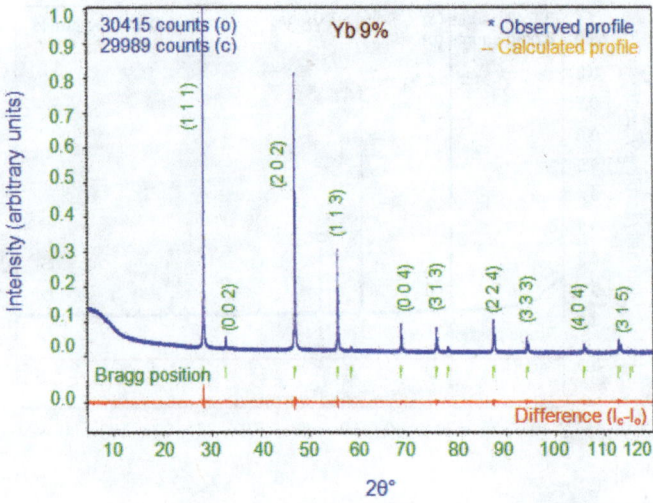

Figure 2.6.2(d) Refined powder XRD profile of $Ca_{1-x}Yb_xF_2$ for $x = 0.09$.

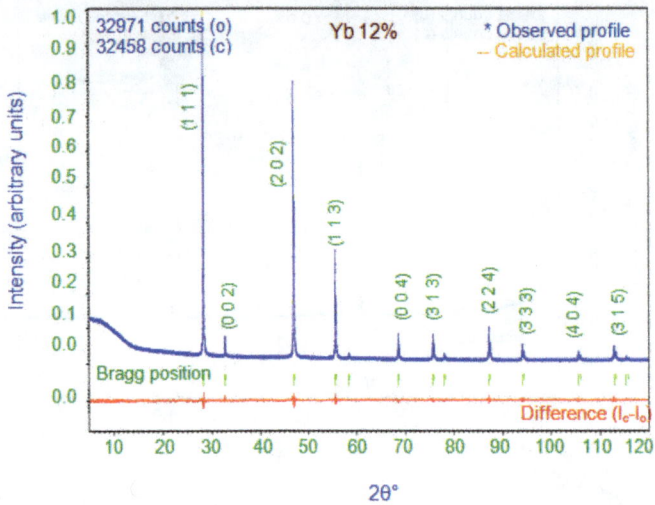

Figure 2.6.2(e) Refined powder XRD profile of $Ca_{1-x}Yb_xF_2$ for $x = 0.12$.

2.7 Aluminium oxide (Al_2O_3)

2.7.1 Powder X-ray data analysis of Al_2O_3, $Cr:Al_2O_3$ and $V:Al_2O_3$

High purity analytical grade Al_2O_3, 5% $Cr:Al_2O_3$ and 5% $V:Al_2O_3$ materials purchased from the scientific suppliers have been used for the present analysis. In order to analyze the bonding and structural behaviour of Al_2O_3, $Cr:Al_2O_3$ and $V:Al_2O_3$, powder X-ray intensity data sets were collected in the 2θ range from $10°$ to $120°$ with step size $0.02°$ at Regional Research Laboratory (RRL), Council of Scientific and Industrial Research (CSIR), Trivandrum, India, using an X'-PERT PRO (Philips, Netherlands), X-ray diffractometer with a monochromatic incident beam of wavelength 1.54056 Å, offering pure CuK_α radiation. A standard software package (Holland and Redfern, 1997) was used to refine the cell parameters of Al_2O_3, $Cr:Al_2O_3$ and $V:Al_2O_3$ using the observed 2θ values. The raw intensity data of Al_2O_3, $Cr:Al_2O_3$ and $V:Al_2O_3$ were refined using Rietveld (1967) refinement. It is a well known powder profile fitting method for the structural refinement. The cell parameters and other structural parameters were refined using this method, by JANA2006 software (Petříček et al., 2006). The structural parameters refined using JANA 2006 (Petříček et al., 2006) is given in tables 2.7.1. The fitted and the observed profiles using JANA 2006 (Petříček et al., 2006) were shown in figures 2.7.1, 2.7.2 and 2.7.3 corresponding to Al_2O_3, $Cr:Al_2O_3$ and $V:Al_2O_3$ respectively. The refined structure factor values of Al_2O_3, $Cr:Al_2O_3$ and $V:Al_2O_3$ are given in the tables 2.7.2, 2.7.3 and 2.7.4 respectively.

Table 2.7.1 Refined structural parameters of Al_2O_3, $Cr:Al_2O_3$ and $V:Al_2O_3$.

Parameter	Al_2O_3	$Cr:Al_2O_3$	$V:Al_2O_3$
a = b (Å)	4.7585(2)	4.7524(1)	4.7585(2)
c(Å)	12.9886(9)	12.9886(9)	12.9886(9)
R_p (%)	8.97	17.35	8.97
$_wR_p$ (%)	14.89	25.55	14.89
R_{obs} (%)	2.04	6.30	2.04
$_wR_{obs}$ (%)	1.77	5.48	1.98
GOF	0.36	0.59	0.44

R_P - Reliability index for profile

$_wR_P$ - Weighted reliability index for profile

R_{obs} - Reliability index for observed structure factors

$_wR_{obs}$ - Weighted reliability index for observed structure factors

GOF - Goodness of Fit

Table 2.7.2 The observed and calculated structure factors of Al_2O_3.

h	k	l	F_{obs}	F_{cal}	$\sigma(F_{obs})$
1	-1	2	48.3383	48.2008	0.8826
1	0	4	84.1760	84.1497	1.2925
2	-1	0	60.6156	62.0532	1.3306
0	0	6	14.2676	12.4067	2.5163
2	-1	3	78.2563	77.3611	1.2108
2	0	2	12.5967	13.6967	2.0585
2	-2	4	95.6387	95.8360	1.9009
2	-1	6	107.2880	107.9540	1.6545
3	-1	1	19.3506	18.1358	1.7443
3	-2	2	20.9373	21.2899	0.7947
1	-1	8	48.7179	49.3864	1.8203
	-1	4	79.9110	80.5196	1.7702
3	0	0	147.9870	145.8890	2.8694
3	-2	5	16.2503	16.8913	2.0981
2	0	8	18.4913	21.3429	3.2945
1	0	10	95.1861	93.6907	2.3402
2	-1	9	47.3512	47.1981	1.1600
3	-1	7	16.4929	16.2219	0.9691
4	-2	0	52.8787	52.4503	3.0424
3	-3	6	10.6940	11.3213	2.6659
3	0	6	13.8605	14.5848	3.4695
4	-2	3	37.3159	40.0770	2.2610
4	-3	1	7.9359	8.1730	0.7932
4	-1	2	32.6067	33.8660	1.7048
3	-2	8	27.7274	28.8782	1.4472
2	-2	10	69.9424	69.5358	3.1005
0	0	12	58.8178	58.9483	2.6731
4	-3	4	51.9194	51.9550	2.1576
4	-1	5	10.6818	10.8671	0.3507
4	-2	6	78.5066	79.7195	2.3307
4	-4	2	32.8624	36.5203	4.1094
3	-1	10	64.6278	65.2878	2.3126
2	-1	12	11.1679	11.4198	1.2323
4	0	4	35.3766	35.7910	3.7224
5	-2	1	9.9304	10.8628	1.9311
3	-2	11	14.6914	16.0182	2.5008
5	-3	2	5.7502	6.0675	0.4137
1	-1	14	57.3517	56.9389	1.8475
5	-1	0	45.0324	45.1706	2.1931
5	-3	5	1.3245	1.2215	0.4563

F_{obs} - observed structure factor

F_{cal} - calculated structure factor

$\sigma(F_{obs})$ - standard error in the measurements

Table 2.7.3 The observed and calculated structure factors of Cr:Al$_2$O$_3$.

h	k	L	F_{obs}	F_{cal}	$\sigma(F_{obs})$
1	-1	2	38.1271	39.0012	0.7112
1	0	4	46.9813	48.1889	1.3925
2	-1	0	22.1570	23.0077	0.6532
0	0	6	26.1386	21.9079	6.0119
2	-1	3	75.0435	74.9115	1.0011
2	0	2	7.8394	8.9825	0.9655
2	-2	4	76.6072	74.1394	1.5028
2	-1	6	76.9439	82.3032	1.5769
3	-1	1	9.9189	7.5395	1.1115
3	-2	2	6.4117	7.1455	0.6238
1	-1	8	36.6175	41.5002	3.6128
3	-1	4	46.1478	47.3026	1.1664
3	0	0	63.4746	63.0466	1.3367
3	-2	5	10.2562	12.4431	1.6570
2	0	8	23.9371	31.0875	3.5907
1	0	10	139.7160	136.8770	4.9119
2	-1	9	52.3583	52.3129	1.8520
3	-1	7	20.5422	23.0860	0.9012
4	-2	0	28.2237	31.7605	1.2411
3	-3	6	7.2633	9.5540	1.6259
3	0	6	8.3649	11.0031	1.8724
4	-2	3	12.6268	13.1945	1.1847
4	-3	1	7.0070	7.1470	0.6380
4	-1	2	6.3310	7.1964	0.5104
3	-2	8	24.7515	26.9183	2.0606
2	-2	10	54.9681	61.5868	5.0839
0	0	12	132.9690	141.3970	5.7655
4	-3	4	28.2734	29.4936	1.2063
4	-1	5	6.9024	7.0983	0.1949
4	-2	6	58.1742	59.8378	1.6247
4	-4	2	7.0981	7.6432	0.8511
3	-1	10	85.2839	85.9777	2.9402
2	-1	12	43.1483	45.5692	1.8961
4	0	4	18.3065	20.1345	0.7610
5	-2	1	0.6329	0.6089	0.0500
3	-2	11	33.1413	34.4737	1.6225
5	-3	2	2.2577	2.4129	0.1218
1	-1	14	11.8463	9.9487	0.7968
5	-1	0	16.1706	11.6982	2.5781
5	-3	5	5.2646	5.5512	0.7650

F_{obs} - observed structure factor

F_{cal} - calculated structure factor

$\sigma(F_{obs})$ - standard error in the measurements

Table 2.7.4 The observed and calculated structure factors of $V:Al_2O_3$.

h	k	l	F_{obs}	F_{cal}	$\sigma(F_{obs})$
1	-1	2	50.4663	50.4664	0.8199
1	0	4	57.5796	57.1930	0.6506
2	-1	0	74.3983	75.7031	1.6537
0	0	6	11.0351	16.7270	1.8978
2	-1	3	57.2229	54.9885	1.2752
2	0	2	4.8111	10.7407	3.2520
2	-2	4	103.0530	105.4350	2.0667
2	-1	6	76.6579	81.0726	1.5441
3	-1	1	16.7901	19.2240	2.0411
3	-2	2	23.7910	25.0217	1.1425
1	-1	8	27.5665	28.9514	0.9163
3	-1	4	110.1920	99.6414	2.3688
3	0	0	145.0820	143.2720	1.8873
3	-2	5	6.7616	6.2949	3.6558
2	0	8	5.6810	1.0605	0.9244
1	0	10	94.8970	92.8043	2.4980
2	-1	9	29.4475	28.5726	1.0022
3	-1	7	16.0118	16.2044	1.5536
4	-2	0	72.6221	69.6345	2.6180
3	-3	6	14.0099	21.8842	3.2515
3	0	6	11.8058	18.4412	2.7399
4	-2	3	59.0789	59.8875	2.9507
4	-3	1	10.0538	11.5755	1.1468
4	-1	2	33.3080	33.8022	2.9261
3	-2	8	12.4249	11.8735	1.0101
2	-2	10	58.9399	55.8785	3.4037
0	0	12	61.9513	56.2399	5.1724
4	-3	4	51.3741	53.1106	2.8530
4	-1	5	12.8961	20.1761	3.2394
4	-2	6	38.7593	39.1171	2.5027
4	-4	2	19.7544	11.9928	2.3015
3	-1	10	42.9167	45.8575	2.6501
2	-1	12	10.1680	11.5640	2.1595
4	0	4	20.5028	30.2358	3.8840
5	-2	1	19.5611	28.5511	3.8977
3	-2	11	3.2966	4.2173	0.5615
5	-3	2	22.8295	23.1090	2.5576
1	-1	14	32.9133	33.7014	1.9853
5	-1	0	80.3483	66.5714	3.6157
5	-3	5	2.3246	2.2432	0.3850

F_{obs} - observed structure factor

F_{cal} - calculated structure factor

$\sigma(F_{obs})$ - standard error in the measurements

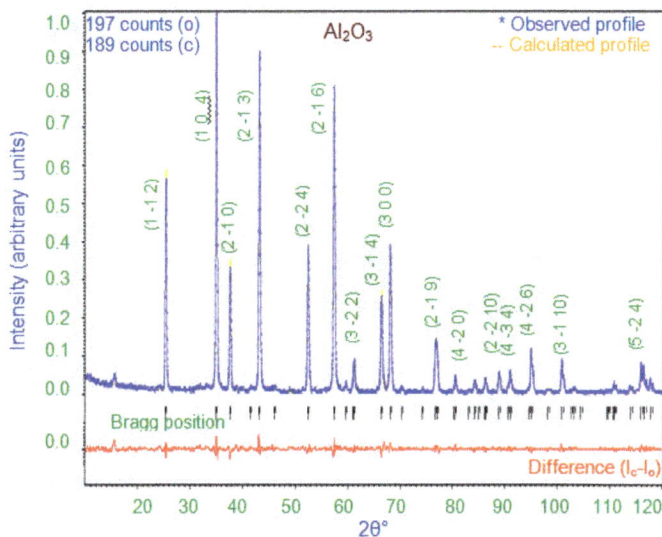

Figure 2.7.1 Refined powder XRD profile of Al$_2$O$_3$.

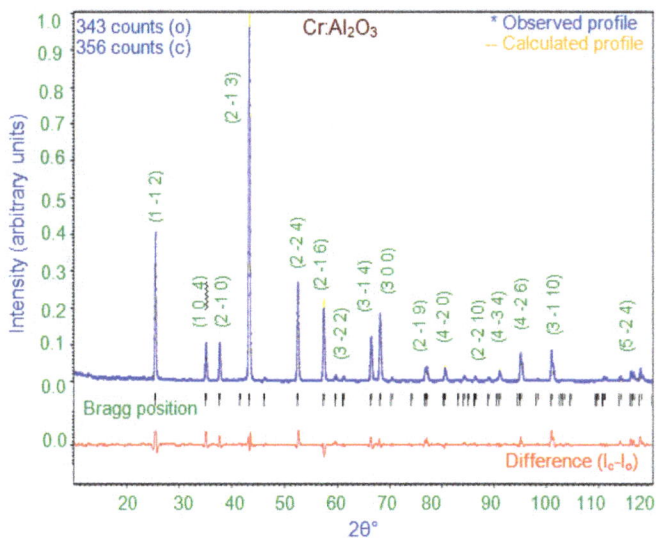

Figure 2.7.2 Refined powder XRD profile of Cr:Al$_2$O$_3$.

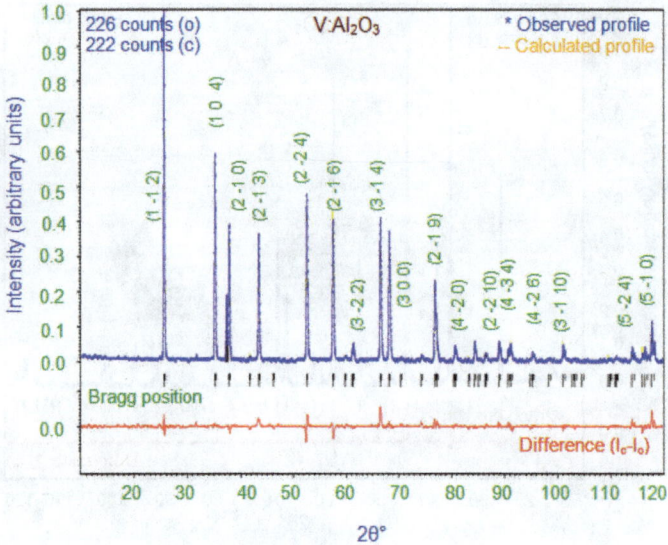

Figure 2.7.3 Refined powder XRD profile of $V:Al_2O_3$.

2.8 Conclusion

The non-linear optical materials of $Ce:Gd_3Ga_5O_{12}$ and $LiNbO_3$ materials were prepared by sol-gel and Czochralski method respectively whereas $Yb:CaF_2$ materials were prepared by co-precipitation method. All the prepared materials were characterized by powder X-ray diffraction technique for analyzing the detailed structural information through JANA2006 software (Petricek et al., 2006). Phase pure systems are observed for all the materials. No additional phases are detected in any of the prepared samples. The structure factors and other structural information have been extracted from Rietveld (1967) refinement technique and further they have been utilized for constructing charge density distribution in the unit cell which is discussed in chapter IV.

References

[1]   Collins D. M., Nature, Vol. 298, pp. 49, 1982. https://doi.org/10.1038/298049a0

[2]   Dollase W. A. J., J. Appl. Cryst., Vol. 19, pp. 267, 1986.
       https://doi.org/10.1107/S0021889886089458

[3]   Gerward L., Olsen J. S., Streenstrup S., Malinowski M., Asbrink S and
        Waskowska A., J. Appl. Crystallogr., Vol. 25, pp. 578, 1992.
        https://doi.org/10.1107/S0021889892004096

[4]   Gull S. F and Daniel, G. J., Nature, Vol. 272, pp. 686, 1978.
        https://doi.org/10.1038/272686a0

[5]   Holland T. J. B and Redfern S. A. T., Mineral. Mag., Vol. 61, pp. 65, 1997.
        https://doi.org/10.1180/minmag.1997.061.404.07

[6]   Howard C. J., J. Appl. Crystallogr., Vol. 15, pp. 615, 1982.
        https://doi.org/10.1107/S0021889882012783

[7]   Izumi F and Dilanian R. A., Recent Research Developments in Physics, Vol. 3,
        Part II, Transworld Research Network, Trivandrum, pp. 699, 2002

[8]   Kar S., Bhatt R., Bartwal K. S and Wadhawan V. K., Cryst. Res. Technol., Vol.
        39, pp. 230, 2004. https://doi.org/10.1002/crat.200310175

[9]   Lyberis A., Stevenson A, J., Suganuma A., Ricaud S., Druon F., Herbst F., Vivien
        D., Gredin P and Mortier M., Optical Materials, Vol. 34, pp. 965, 2001.
        https://doi.org/10.1016/j.optmat.2011.05.036

[10]   March A., Z. Kristallogr., Vol. 81, pp. 285, 1932

[11]   Thompson P., Cox D. E and Hastings J. B., J. Appl. Cryst., Vol. 20, pp. 79, 1987.
        https://doi.org/10.1107/S0021889887087090

[12]   Petříček V., Dušek M and Palatinus L., JANA 2006, The crystallographic
        computing system, Institute of Physics, Academy of sciences of the Czech
        republic, Praha, 2006

[13]   Rietveld H. M. J. Appl. Crystallogr., Vol. 2, pp. 65, 1969.
        https://doi.org/10.1107/S0021889869006558

[14]   Shannon R. D., Acta Crystallographica, Vol. A32, pp. 751, 1976.
        https://doi.org/10.1107/S0567739476001551

Chapter 3

Optical and Size Analysis of
Non-Linear Optical Materials

Abstract

Chapter III deals with the optical properties and micro-structural characterization of non-linear optical materials such as $PbMoO_4$, $LiNbO_3$, $Ce:Gd_3Ga_5O_{12}$, $CaCO_3$, $Yb:CaF_2$, and Al_2O_3, $Cr:Al_2O_3$,$V:Al_2O_3$ in a detailed manner. The band gap, the crystallite size and the particle size of the chosen non-linear optical materials from UV-visible analysis, powder X-ray profile and scanning electron microscope respectively are also determined and discussed in this chapter. This chapter also discusses about elemental compositional analysis for $PbMoO_4$, $LiNbO_3$, $Ce:Gd_3Ga_5O_{12}$, $CaCO_3$, $Yb:CaF_2$, and Al_2O_3, $Cr:Al_2O_3$,$V:Al_2O_3$.

Keywords

Particle Size, Grain Size, Optical Band Gap, Energy Dispersive X-Ray Analysis, Scanning Electron Microscopy

Contents

3.1   Introduction

3.1.1   Optical band gap analysis

The interaction of electromagnetic radiation having a suitable wavelength with the material is much more useful for understanding the mechanisms responsible for their optical behaviour. Depending on the nature of the material and also on the characteristics of incident light, many optical phenomena (Pancove, 1971) occur such as absorption, transmission, reflectance, etc. Optical absorption, electrical conductivity and refractive index are some of the properties of the materials that depend on the optical band gap. The spectral absorption coefficient has been evaluated as $\alpha = 4\pi k/\lambda$, where k is the spectral extinction coefficient. Near the band gap, absorption coefficient (α) steeply increases with increasing photon energy. As a result, the absorption edge is indicative of the location of the band gap. Accurate estimation of the band gap requires use of the following formula (Wood and Tauc, 1972)

$$(\alpha E)^n = B(h\nu - E_g) \tag{3.1}$$

where α is the optical absorption coefficient, E is photon energy, B is a constant depending on the transition probability, E_g is the band gap energy and n is an index that characterizes the optical absorption process and it is considered as 1/2, 2, 1/3, and 2/3 for indirect allowed, direct allowed, indirect forbidden and direct forbidden respectively. In the present work, optical property such as the band gap of some non-linear optical materials has been analyzed through UV-Visible spectroscopy.

3.1.2 Size and energy dispersive X-ray spectroscopic analysis

In order to understand the properties of the material it is required to examine the particle/grain size and elemental compositions. Therefore, the characterization techniques such as X-ray crystallography need to be combined with local examination using conventional electron microscopy techniques such as imaging and energy dispersive X-ray spectroscopy.

Various research groups have reported different methods to measure the particle size such as dynamic light scattering (Berne and Pecora, 2000), sedimentation (Hideto et al., 2001), image analysis (Janaka et al., 2012), laser diffraction (Boer et al., 1987) and acoustic spectroscopy (Alba et al., 1999) etc.

Scanning electron microscopy (SEM) (McMullan, 2006) is a powerful characterization technique used for the analysis of morphological properties and size distribution of the powder samples using imaging methods. Scanning electron microscope (SEM) can be utilized for high magnification of almost all materials. In the present work, the size analysis of the particles of the powder samples has been done for the chosen non-linear optical materials through scanning electron microscopy technique.

The grain size of the chosen non-linear optical material was evaluated using Scherrer's formula (Scherrer, 1918) which employs GRAIN software formulated by Saravanan (Saravanan, Private communication). The size of the chosen non-linear optical material was analyzed using full width at half maximum of the powder XRD peaks.

$$D_V = \frac{k\lambda}{\beta \cos \theta} \tag{3.2}$$

where, D_V is the crystallite size (size of the coherently diffracting domain), k is a constant - usually 0.9 when the particle is in the spherical nature, λ is the wavelength used, β is FWHM (Full Width at Half Maximum) in radians, and θ is the Bragg angle of the reflection.

Energy Dispersive X-ray spectroscopy (EDS or EDAX) (Goldstein, 2003), is an analytical technique used to identify the elemental composition of a samples. Another way to use EDS with SEM is to make a quantitative chemical analysis of the material. Also, it provides information on the purity of the powder samples. In the present work, energy dispersive X-ray spectroscopic (EDS) analysis has been done for some non-linear optical materials.

3.2   Lead molybdate (PbMoO$_4$)

3.2.1   Optical band gap analysis of PbMoO$_4$

In the present study, lead molybdate (PbMoO$_4$) sample was characterized by UV-Visible spectrophotometer. Figure 3.2.1.1 shows the dependence of $(\alpha E)^2$ on photon energy (E) for PbMoO$_4$. The band gap (E_g) is the intercept of the straight line obtained by plotting $(\alpha E)^2$ vs E. The optical band gap energy (E_g) of PbMoO$_4$ was evaluated (Wood and Tauc, 1972) as given by the equation (3.1). The energy gap values are given in table 3.2.1.1. The value of band gap energy determined for PbMoO$_4$ is 3.13 eV. The band gap energy determined in the present work was found to agree well with the other research workers (Sczancoski et al., 2009; Zhang et al., 1998).

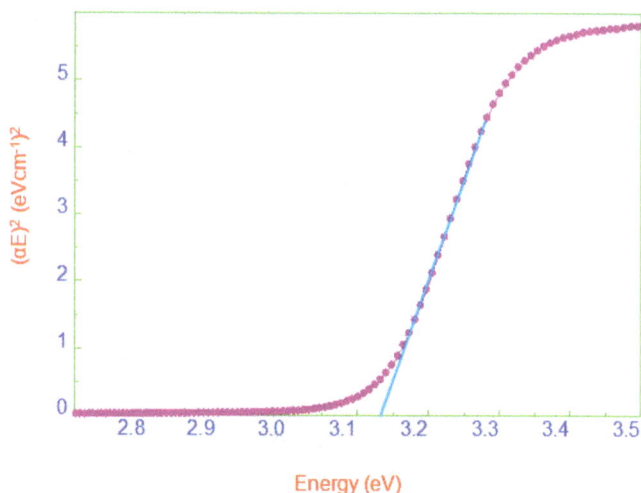

Figure 3.2.1.1 Dependence of $(\alpha E)^2$ on photon energy for PbMoO$_4$.

Table 3.2.1.1 Energy gap values of PbMoO$_4$.

Reference	Energy gap (eV)
(Sczancoski et al., 2009)	3.14 – 3.19
(Zhang et al., 1998)	2.59 – 3.62
Present work	3.13

3.2.2 Size and EDS analysis of PbMoO$_4$

The SEM micrograph of PbMoO$_4$ is shown in figure 3.2.2.1. The crystallite size of PbMoO$_4$ was evaluated using the software GRAIN formulated by Saravanan (Saravanan, private communication). The crystallite size of PbMoO$_4$ is analyzed using full width at half maximum (FWHM) of the powder XRD peaks. From this analysis, the crystallite size (r_{Xray}) comes out to be ~31 nm. The particle size (r_{SEM}) from SEM measurement is ~2841 nm. The number of coherently diffracting domains can be obtained using the equation

$$N = \frac{r_{SEM}}{r_{Xray}} \tag{3.3}$$

Hence, there are approximately 91 coherently diffracting domains in each particle (Guinier, 1994). The crystallite size of PbMoO$_4$ is evaluated by the powder XRD gives only the size of the coherently diffracting domains, it cannot be directly compared to the size obtained using an SEM except in rare cases (Scherrer, 1918).

The presence of appropriate quantities of Pb, Mo and O species in PbMoO$_4$ has been revealed through the EDS spectrum as shown in figure 3.2.2.2. No additional impurity was detected in the EDS spectrum of PbMoO$_4$.

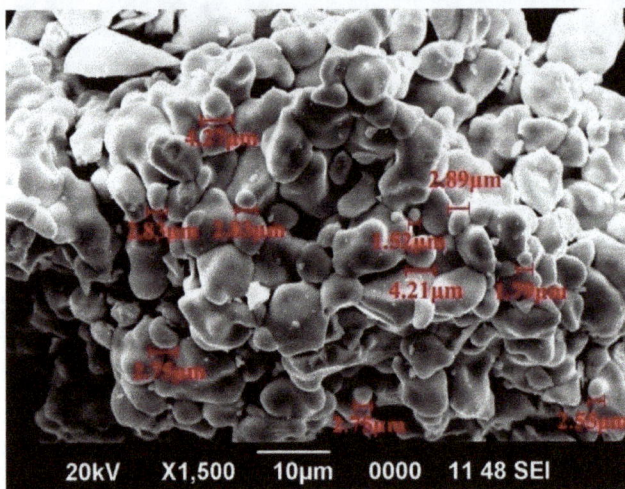

Figure 3.2.2.1 SEM micrograph of PbMoO$_4$.

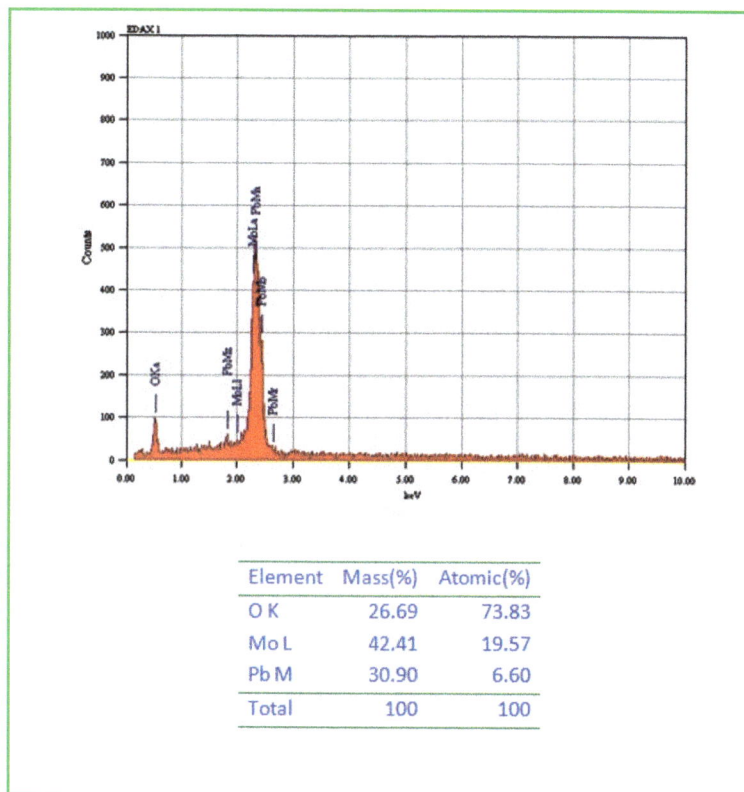

Element	Mass(%)	Atomic(%)
O K	26.69	73.83
Mo L	42.41	19.57
Pb M	30.90	6.60
Total	100	100

Figure 3.2.2.2 EDS picture of PbMoO₄.

3.3   Lithium niobate (LiNbO₃)

3.3.1   Optical band gap analysis of LiNbO₃

UV–Visible spectroscopy is one of the important methods to understand the optical properties of the materials. The optical band gap energy (E_g) was determined by using the relation given by the equation (3.1) (Wood and Tauc, 1972). Figure 3.3.1.1 shows linear fitting to the $(\alpha E)^2$ against energy (E) plot for LiNbO₃. The band gap energy for LiNbO₃ was determined and tabulated along with the previously reported values in table 3.3.1.1 (Rajeev Bhatt et al., 2003; Jaejun et al., 1997; Kityk et al., 2001; Dhar and

Mansingh, 1990). In the present work, the band gap energy agrees well with the already reported values.

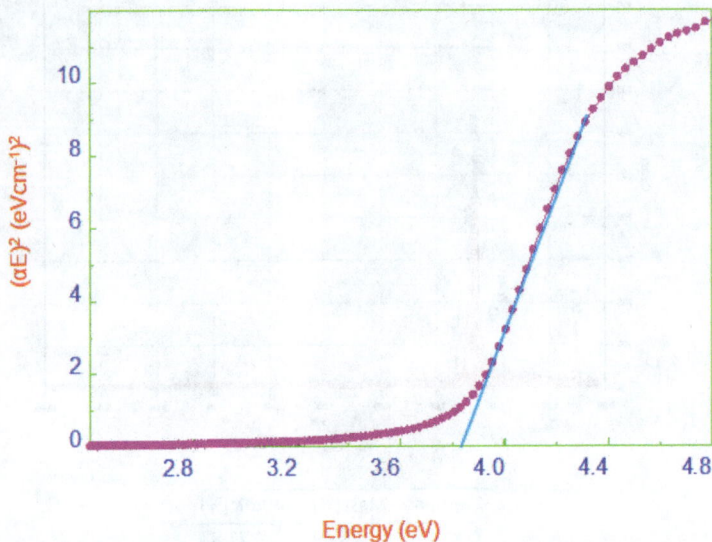

Figure 3.3.1.1 Dependence of $(\alpha E)^2$ on photon energy of LiNbO₃.

Table 3.3.1.1 Energy gap values of LiNbO₃.

Reference	Energy gap (eV)
(Rajeev Bhatt et al., 2003)	3.79
(Jaejun et al., 1997)	3.70
(Kityk et al., 2001)	3.69
(Dhar and Mansingh, 1990)	3.78
Present work	3.81

3.3.2 Size and EDS analysis of LiNbO$_3$

Figure 3.3.2.1 shows the SEM micrograph of LiNbO$_3$ with a magnification of 3000. The crystallite size of LiNbO$_3$ was evaluated using the software GRAIN formulated by Saravanan (Saravanan, Private communication). The crystallite size of LiNbO$_3$ was analyzed using full width at half maximum (FWHM) of the powder XRD peaks. From this analysis, the crystallite size (r_{Xray}) is found to be ~1.9 nm. The particle size (r_{SEM}) from SEM measurement comes out to ~1108 nm. The crystallite size can be obtained using the equation (3.2). The crystallite size is the size of the coherently diffracting domains. Hence, there are approximately 583 coherently diffracting domains in each particle (Guinier, 1994). The crystallite size of LiNbO$_3$ is evaluated by powder XRD which can't be directly compared with the particle size derived from SEM micrograph except in rare cases (Scherrer, 1918).

The EDS spectrum in figure 3.3.2.2 shows the presence of appropriate quantities of the constituent elements of LiNbO$_3$. No additional impurity was detected in the EDS spectrum of LiNbO$_3$.

Figure 3.3.2.1 SEM micrograph of LiNbO$_3$.

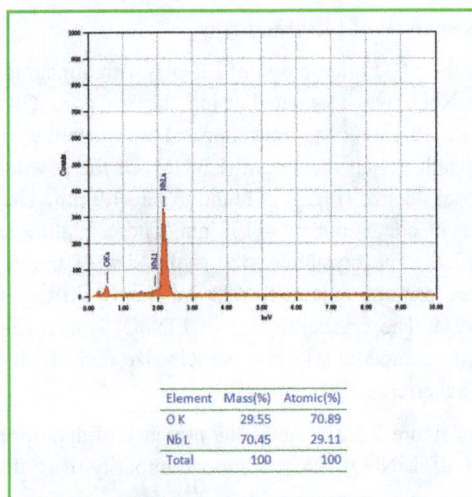

Figure 3.3.2.2 EDS picture of LiNbO$_3$.

3.4 Gadolinium gallium garnet (Gd$_3$Ga$_5$O$_{12}$)

3.4.1 Optical band gap analysis of Gd$_{3-x}$Ce$_x$Ga$_5$O$_{12}$

UV-Visible spectroscopy is ideal for characterizing optical and electronic properties of different materials. The prepared non-linear optical material Gd$_{3-x}$Ce$_x$Ga$_5$O$_{12}$ (x = 0.5, 1 and 3) was characterized using a UV-Visible spectrophotometer (Cary 5000 - Varian) at SAIF, CUSAT, Cochin, India, to estimate band gap energy (E_g). Wood and Tauc (1972) have related the energy of the incident photon to the band gap as given in equation (3.1). According to these authors, the band gap energy (E_g) of Gd$_{3-x}$Ce$_x$Ga$_5$O$_{12}$ (x = 0.5, 1 and 3) was estimated by fitting $(\alpha E)^2$ against energy(E)and presented in the table 3.4.1.1. The dependence of $(\alpha E)^2$ on photon energy (E) of Gd$_{3-x}$Ce$_x$Ga$_5$O$_{12}$ (x = 0.5, 1 and 3) is shown in figures 3.4.1.1 - 3.4.1.3 respectively.

Table 3.4.1.1 Energy gap values of Gd$_{3-x}$Ce$_x$Ga$_5$O$_{12}$.

Composition of Ce (%)	Energy gap (eV)
0.5	5.40
1	5.42
3	5.39

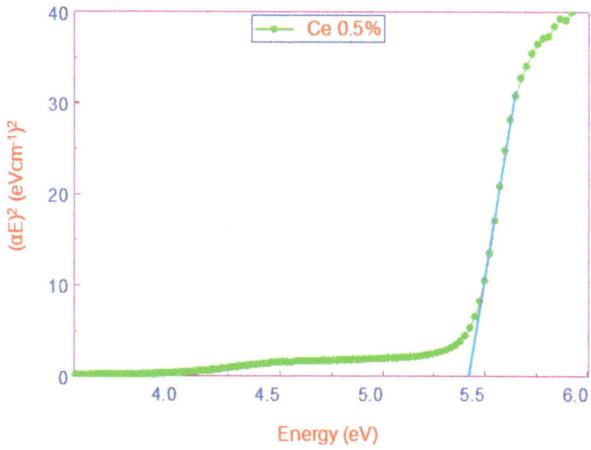

Figure 3.4.1.1 Dependence of $(\alpha E)^2$ on photon energy of $Gd_{3-x}Ce_xGa_5O_{12}$ for $x = 0.5$.

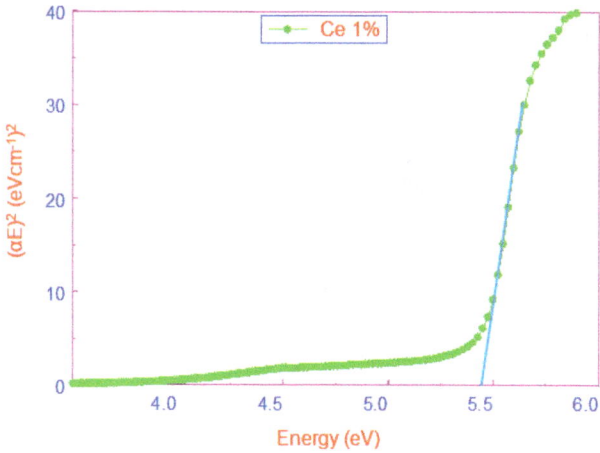

Figure 3.4.1.2 Dependence of $(\alpha E)^2$ on photon energy of $Gd_{3-x}Ce_xGa_5O_{12}$ for $x = 1$.

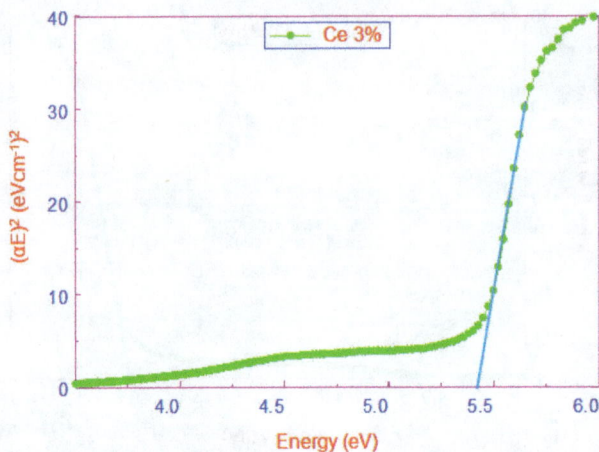

Figure 3.4.1.3 Dependence of $(\alpha E)^2$ on photon energy of $Gd_{3-x}Ce_xGa_5O_{12}$ for $x = 3$.

3.4.2 Size analysis of $Gd_{3-x}Ce_xGa_5O_{12}$

The synthesized non-linear optical material $Gd_{3-x}Ce_xGa_5O_{12}$ (x = 0.5, 1 and 3) were characterized by scanning electron microscope (SEM). Figures 3.4.2.1 - 3.4.2.3 show typical scanning electron microscopy (SEM) micrographs of $Gd_{3-x}Ce_xGa_5O_{12}$ (x = 0.5, 1 and 3) respectively. The crystallite size of the $Gd_{3-x}Ce_xGa_5O_{12}$ (x = 0.5, 1 and 3) was analyzed using full width at half maximum (FWHM) of the powder XRD peaks using the software GRAIN software formulated by Saravanan (Saravanan, private communication). The estimated values of crystallite sizes and particle sizes of $Gd_{3-x}Ce_xGa_5O_{12}$ (x = 0.5, 1 and 3) are summarized in the table 3.4.2.1.

Table 3.4.2.1 The crystallite and particle size of $Gd_{3-x}Ce_xGa_5O_{12}$ (x = 0.5, 1 and 3).

Compositions of Yb (%)	Crystallite size from XRD (nm)	Average particle size from SEM (nm)
0.5	252	451
1	154	263
3	97	50

Figure 3.4.2.1 SEM micrograph of $Gd_{3-x}Ce_xGa_5O_{12}$ for x = 0.5.

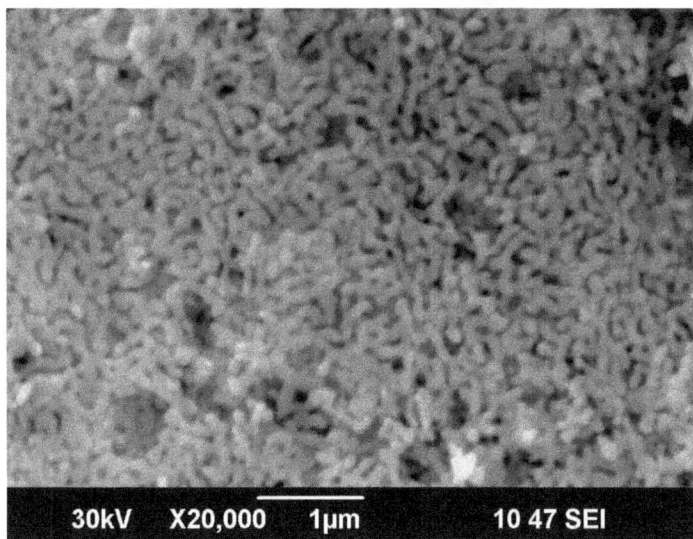

Figure 3.4.2.2 SEM micrograph of $Gd_{3-x}Ce_xGa_5O_{12}$ for x = 1.

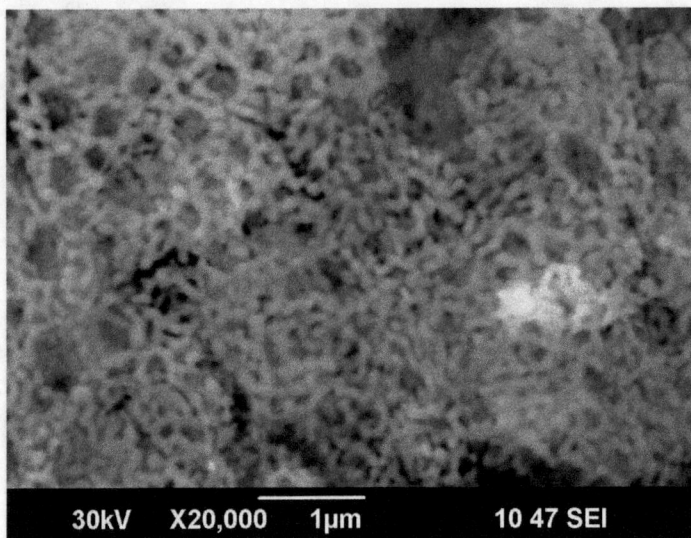

Figure 3.4.2.3 SEM micrograph of $Gd_{3-x}Ce_xGa_5O_{12}$ for x = 3.

3.5 Calcite (CaCO₃)

3.5.1 Optical band gap analysis of CaCO₃

The optical band gap energy(E_g) can be calculated via UV-Visible absorption spectrum using Wood and Tauc plots (Wood and Tauc, 1972). The band gap of the calcite was estimated by plotting$(\alpha E)^2$ vs energy(E) and extrapolating the linear portion near the onset of absorption edge to the energy (E) axis and is shown in figure 3.5.1.1. In the present work, the value of band gap energy of calcite was found to be 3.99 eV and it agrees well with the value reported earlier 3.93 eV (Aydinol et al., 2007).

3.5.2 Size and EDS analysis of CaCO₃

Scanning electron microscope (SEM) micrograph of calcite was obtained under different magnifications on a field emission SEM apparatus (JSM-6390LV, JEOL) operating at acceleration voltage of 30 kV, one of which is shown in figure 3.5.2.1 The particle size of calcite from SEM measurement is found to be ~40 nm. The crystallite size from XRD of calcite was estimated to be about ~13 nm. Hence there are approximately ~3 coherently diffracting domains in each particle of calcite (Guinier, 1994).

The EDS spectrum of calcite in figure 3.5.2.2 shows the presence of appropriate quantities of the constituent elements of calcite. No additional impurity was detected in the EDS spectrum of calcite.

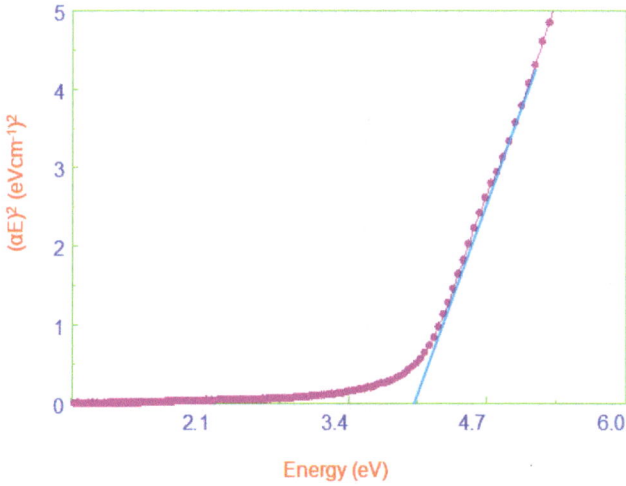

Figure 3.5.1.1 Dependence of $(\alpha E)^2$ on photon energy of CaCO$_3$.

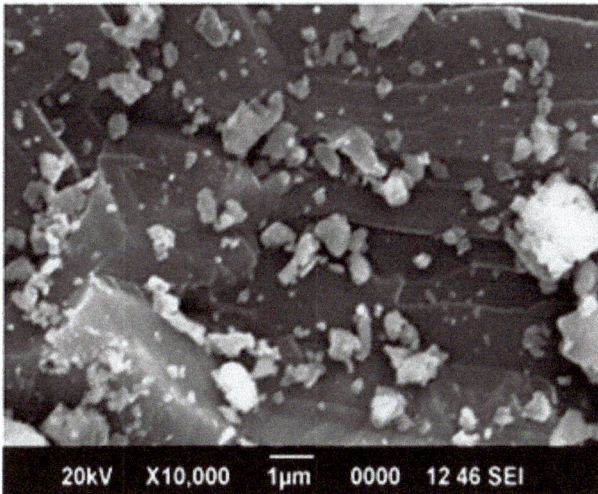

Figure 3.5.2.1 SEM micrograph of CaCO$_3$.

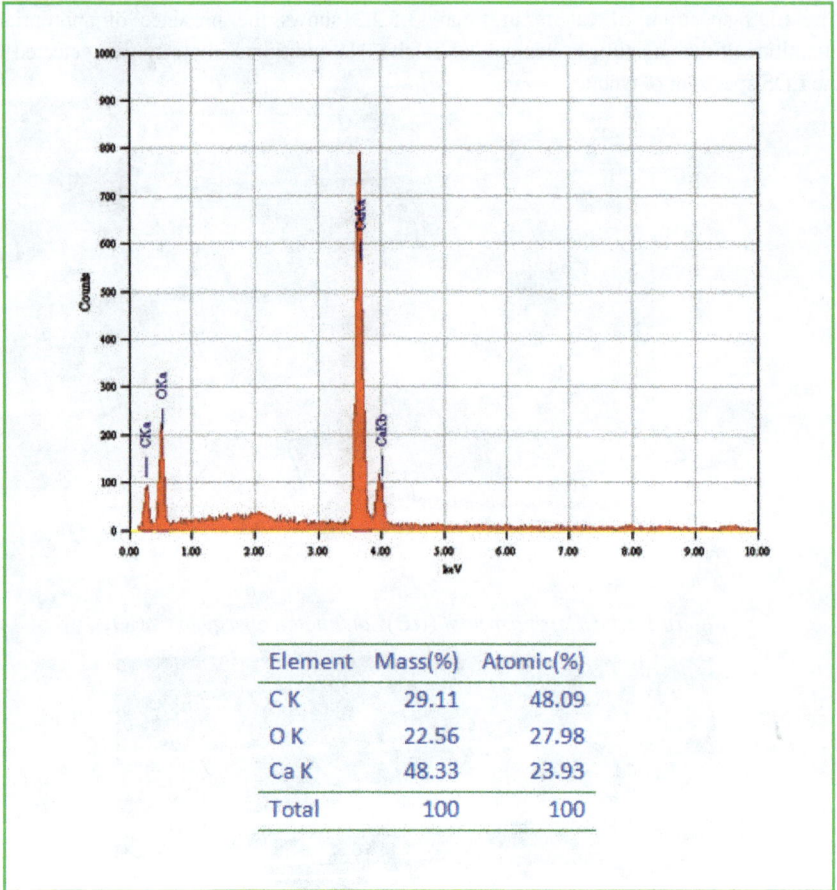

Element	Mass(%)	Atomic(%)
C K	29.11	48.09
O K	22.56	27.98
Ca K	48.33	23.93
Total	100	100

Figure 3.5.2.2 EDS picture of CaCO₃.

3.6 Calcium fluoride (CaF₂)

3.6.1 Size and EDS analysis of Ca₁₋ₓYbₓF₂

The prepared $Ca_{1-x}Yb_xF_2$ (x = 0.00, 0.03, 0.06, 0.09 and 0.12) non-linear optical material was characterized by scanning electron microscope (SEM) with the magnification of

30,000 on SEM instrument (JSM-6390LV, JEOL) operating at acceleration voltage of 20 kV. The SEM micrographs of $Ca_{1-x}Yb_xF_2$ (x = 0.00, 0.03, 0.06, 0.09 and 0.12) materials are shown in figures 3.6.1.1 - 3.6.1.5 respectively. The SEM image corresponding to $Ca_{1-x}Yb_xF_2$ (x= 0.06) shows the formation of clusters inside the matrix and the particle size seems to be large compared to other compositions. The powder XRD gives only the size of the coherently diffracting domain and it is known as crystallite size. The crystallite sizes of the prepared samples were also found using XRD powder data using GRAIN software (Saravanan, private communication) which employs Scherrer formula (Scherrer, 1918). The average particle size (r_{SEM}) and the crystallite size (r_{Xray}) of the prepared $Ca_{1-x}Yb_xF_2$ (x = 0.00, 0.03, 0.06, 0.09 and 0.12) was given in table 3.6.1.1.

Table 3.6.1.1 The crystallite and average particle sizes of $Ca_{1-x}Yb_xF_2$.

Compositions of Yb (%)	Crystallite size from XRD (nm)	Average particle size from SEM (nm)
0	73	200
3	98	110
6	150	270
9	56	80
12	60	140

Figure 3.6.1.1 SEM micrograph of $Ca_{1-x}Yb_xF_2$ for x = 0.00.

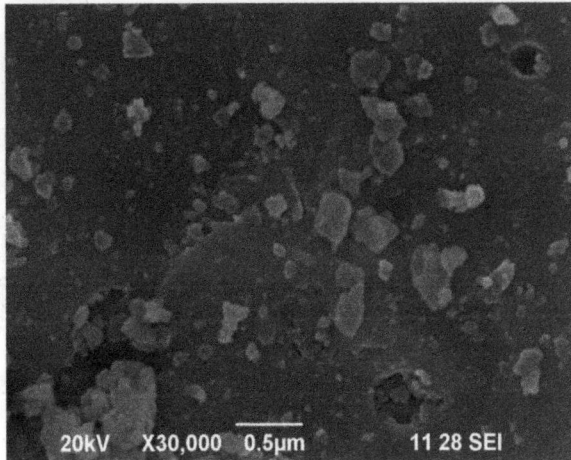

Figure 3.6.1.2 SEM micrograph of $Ca_{1-x}Yb_xF_2$ for x = 0.03.

Figure 3.6.1.3 SEM micrograph of $Ca_{1-x}Yb_xF_2$ for x = 0.06.

Figure 3.6.1.4 SEM micrograph of $Ca_{1-x}Yb_xF_2$ for x = 0.09.

Figure 3.6.1.5 SEM micrograph of $Ca_{1-x}Yb_xF_2$ for x = 0.12.

Energy dispersive X-ray spectroscopy (EDS) measurement of the synthesized $Ca_{1-x}Yb_xF_2$ (x = 0.00, 0.03, 0.06, 0.09 and 0.12) materials are shown in figures 3.6.1.6 - 3.6.1.10 respectively. The peaks in the EDS spectrum of $Ca_{1-x}Yb_xF_2$ (x = 0.00, 0.03, 0.06, 0.09 and 0.12) reveal the presence of appropriate quantities of Ca, F and Yb species respectively. Also, the values of each element present in the prepared samples are tabulated in the EDS spectrum. From the observed spectrum, it is seen that the Yb concentration increases monotonically with decreasing of Ca atom. No additional impurities are identified in the EDS spectrum of $Ca_{1-x}Yb_xF_2$ (x = 0, 0.03, 0.06, 0.09 and 0.12), which implies that the incorporated Yb atoms occupy their preferential sites in the Ca lattice sites.

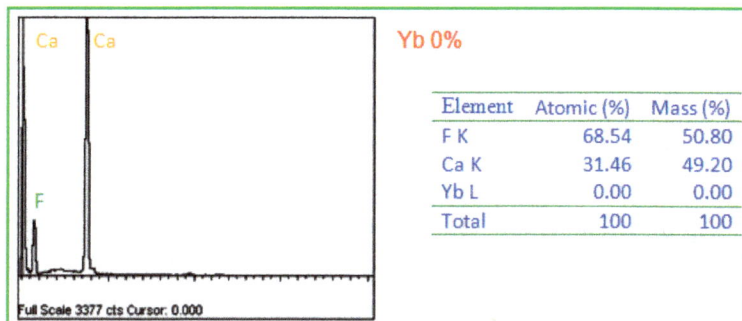

Element	Atomic (%)	Mass (%)
F K	68.54	50.80
Ca K	31.46	49.20
Yb L	0.00	0.00
Total	100	100

Figure 3.6.1.6 EDS picture of $Ca_{1-x}Yb_xF_2$ for x = 0.00.

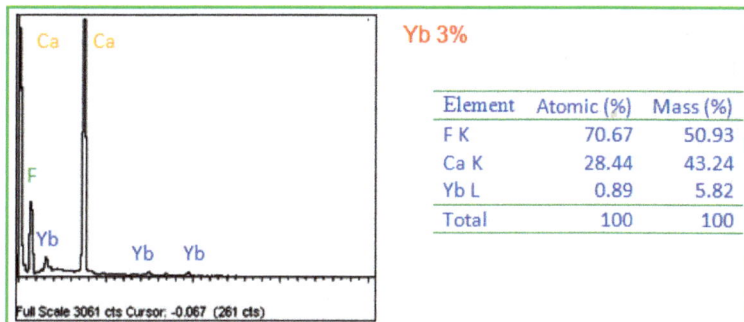

Element	Atomic (%)	Mass (%)
F K	70.67	50.93
Ca K	28.44	43.24
Yb L	0.89	5.82
Total	100	100

Figure 3.6.1.7 EDS picture of $Ca_{1-x}Yb_xF_2$ for x = 0.03.

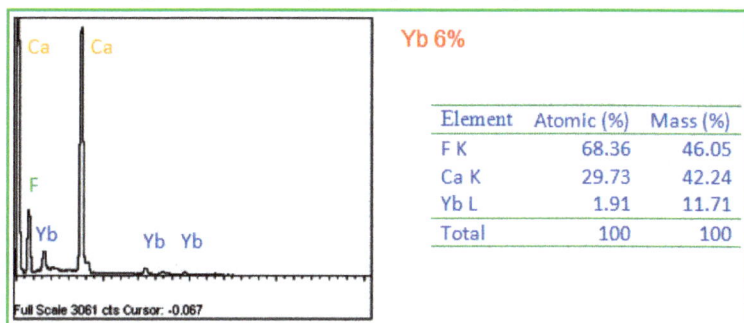

Element	Atomic (%)	Mass (%)
F K	68.36	46.05
Ca K	29.73	42.24
Yb L	1.91	11.71
Total	100	100

Figure 3.6.1.8 EDS picture of $Ca_{1-x}Yb_xF_2$ for x = 0.06.

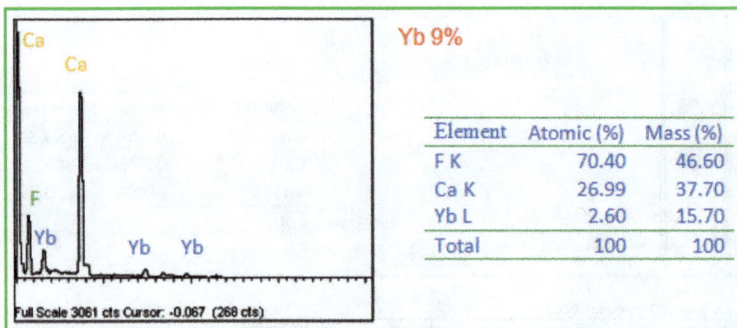

Element	Atomic (%)	Mass (%)
F K	70.40	46.60
Ca K	26.99	37.70
Yb L	2.60	15.70
Total	100	100

Yb 9%

Full Scale 3061 cts Cursor: -0.067 (268 cts)

Figure 3.6.1.9 EDS picture of $Ca_{1-x}Yb_xF_2$ for x = 0.09.

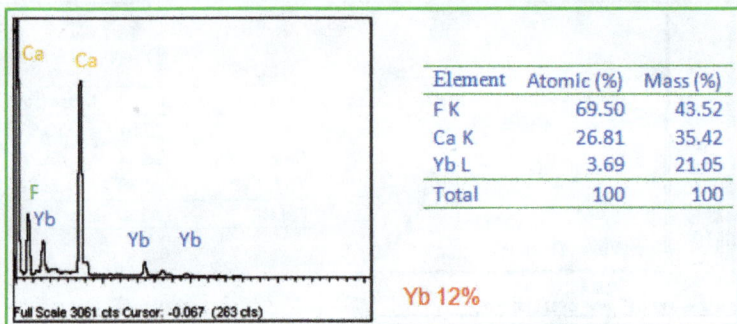

Element	Atomic (%)	Mass (%)
F K	69.50	43.52
Ca K	26.81	35.42
Yb L	3.69	21.05
Total	100	100

Yb 12%

Full Scale 3061 cts Cursor: -0.067 (263 cts)

Figure 3.6.1.10 EDS picture of $Ca_{1-x}Yb_xF_2$ for x = 0.12.

3.7 Aluminium oxide (Al_2O_3)

3.7.1 Size and EDS analysis of Al_2O_3, $Cr:Al_2O_3$ and $V:Al_2O_3$

The scanning electron microscopic (SEM) studies were done for the prepared Al_2O_3, $Cr:Al_2O_3$ and $V:Al_2O_3$ samples with the SEM instrument (JSM-6390LV, JEOL) operating at acceleration voltage of 25 kV. SEM analysis was carried out under different magnifications one of which for Al_2O_3, $Cr:Al_2O_3$ and $V:Al_2O_3$ are shown in figures 3.7.1.1 - 3.7.1.3 respectively.

Figure 3.7.1.1 SEM micrograph of Al$_2$O$_3$.

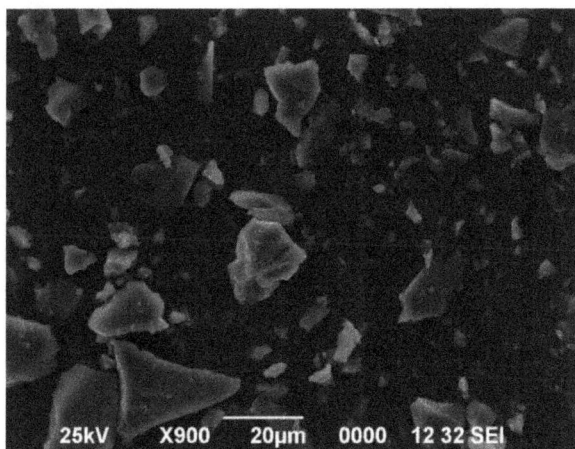

Figure 3.7.1.2 SEM micrograph of Cr:Al$_2$O$_3$.

Figure 3.7.1.3 SEM micrograph of V:Al$_2$O$_3$.

The crystallite sizes of the prepared Al$_2$O$_3$, Cr:Al$_2$O$_3$ and V:Al$_2$O$_3$ samples were determined using powder X-ray diffraction data sets using Scherrer's formula (Scherrer, 1918) which employs GRAIN written by Saravanan (Saravanan, private communication). The average particle size and crystallite size of Al$_2$O$_3$, Cr:Al$_2$O$_3$ and V:Al$_2$O$_3$ estimated was given in the table 3.7.1.1. Hence, there are approximately 8, 12 and 26 domains in Al$_2$O$_3$, Cr:Al$_2$O$_3$ and V:Al$_2$O$_3$ respectively (Guinier, 1994).

The EDS spectrum of Al$_2$O$_3$, Cr:Al$_2$O$_3$ and V:Al$_2$O$_3$ in figure 3.7.1.4 shows the presence of appropriate quantities of the constituent elements. The atomic and mass percentages of the constituent elements of Al$_2$O$_3$, Cr:Al$_2$O$_3$ and V:Al$_2$O$_3$ are also given in the inset of figure 3.7.1.4. No additional impurities are detected in the respective EDS spectrum of Al$_2$O$_3$, Cr:Al$_2$O$_3$ and V:Al$_2$O$_3$.

Table 3.7.1.1 The crystallite sizes and average particle sizes of Al$_2$O$_3$, Cr:Al$_2$O$_3$ and V:Al$_2$O$_3$.

System	Crystallite size from XRD (nm)	Average particle size from SEM (nm)
Al$_2$O$_3$	130	1037
Cr:Al$_2$O$_3$	144	1734
V:Al$_2$O$_3$	133	3441

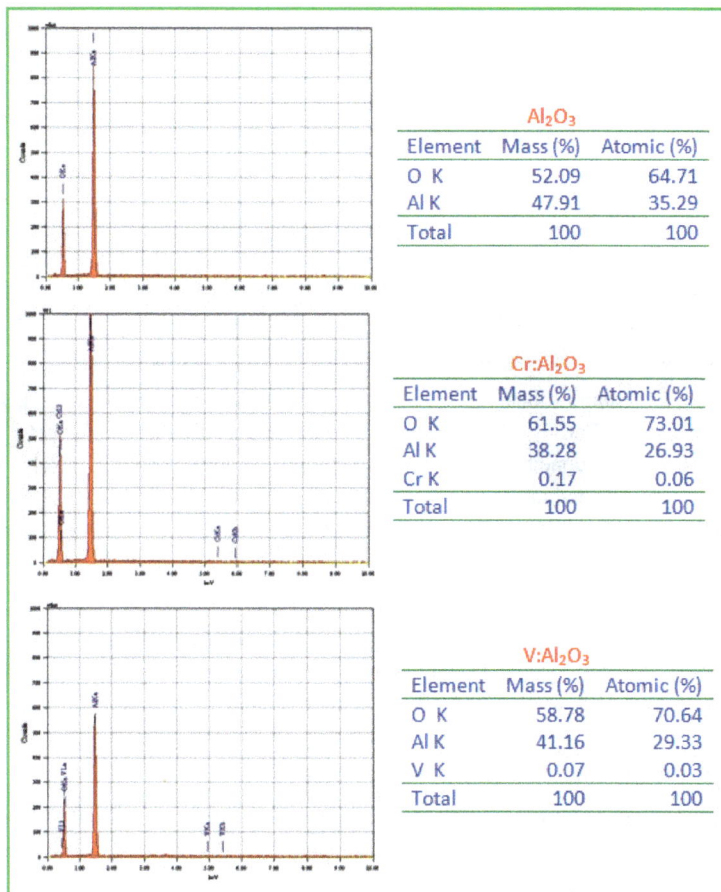

Al_2O_3		
Element	Mass (%)	Atomic (%)
O K	52.09	64.71
Al K	47.91	35.29
Total	100	100

$Cr:Al_2O_3$		
Element	Mass (%)	Atomic (%)
O K	61.55	73.01
Al K	38.28	26.93
Cr K	0.17	0.06
Total	100	100

$V:Al_2O_3$		
Element	Mass (%)	Atomic (%)
O K	58.78	70.64
Al K	41.16	29.33
V K	0.07	0.03
Total	100	100

Figure 3.7.1.4 EDS spectrum of Al_2O_3, $Cr:Al_2O_3$ and $V:Al_2O_3$.

3.8   Conclusion

The chosen non-linear optical materials $PbMoO_4$, $LiNbO_3$, Ce doped $Gd_3Ga_5O_{12}$, and calcite have been characterized by UV-Visible spectrophotometer. Based on this technique, the optical band gap energy (E_g) of the non-linear optical materials have been estimated. The optical band gap energies determined in the present work are found to agree well with the earlier reported values.

The synthesized non-linear optical materials Ce doped $Gd_3Ga_5O_{12}$, $LiNbO_3$, calcite, Yb doped CaF_2, Al_2O_3, Al_2O_3 doped with Cr and V separately and $PbMoO_4$ have been characterized by scanning electron microscopy (SEM) technique. The observed SEM micrographs give the particle sizes of the synthesized non-linear optical materials. The SEM micrographs of all the prepared non-linear optical materials reveal that the particle sizes are in the nano range. The crystallite sizes of the prepared non-linear optical materials have also been estimated using powder X-ray diffraction (XRD) data sets. Both the particle size and crystallite size are compared and the number of coherently diffracting domains is calculated for the prepared non-linear optical materials.

The elemental compositional analysis of the prepared non-linear optical materials, $PbMoO_4$, $LiNbO_3$, $CaCO_3$, Yb doped CaF_2, and Al_2O_3, Al_2O_3 doped with Cr and V separately have been done by energy dispersive X-ray spectroscopy (EDS). No additional impurities were detected in the respective EDS spectrum of the chosen non-linear optical materials.

References

[1]   Alba F., Crawley G. M., Fatkin J., Higgs D. M. J and Kippax P. G., Colloids and Surfaces A: Physicochemical and Engineering Aspects, Vol. 153(1–3), pp. 495, 1999. https://doi.org/10.1016/S0927-7757(98)00473-7

[2]   Aydinol M. K., Mantese J. V and Alpay S. P., J. Phys. Condens. Matter, Vol. 19, pp. 496210, 2007. https://doi.org/10.1088/0953-8984/19/49/496210

[3]   Berne B. J and Pecora R., Dynamic Light Scattering, Courier Dover Publications, 2000

[4]   Boer G. B. J., de Weerd C., Thoenes D and Goossens H. W, J., Particle Characterization, Vol. 4, pp. 138, 1987

[5]   Dhar A and Mansingh A., J. Appl. Phys., Vol. 68, pp. 5804, 1990. https://doi.org/10.1063/1.346951

[6]   Goldstein J., Scanning Electron Microscopy and X-Ray Microanalysis, Springer, 2003. https://doi.org/10.1007/978-1-4615-0215-9

[7]   Guinier A., X-ray Diffraction, Dover Publications, New York, 1994

[8]   Hideto Y., Hiroaki M., Kunihiro F and Yusuke T., Advanced Powder Technology, Vol. 12(1), pp. 79, 2001. https://doi.org/10.1163/156855201744976

[9]   Jaejun Yu and Key-Taeck Park, Physica B, Vol. 237, pp. 341, 1997.
      https://doi.org/10.1016/S0921-4526(97)00210-X

[10] Janaka K., Kimitoshi H and Keita O., International Journal of GEOMATE, Vol. 3,
      pp. 290, 2012

[11] Kityk I. V., Makowska-Janusik M., Fontana M. D., Aillerie M and Abdi F., J.
      Appl. Phys., Vol. 90, pp. 5542, 2001. https://doi.org/10.1063/1.1413942

[12] McMullan D., The Journal of Scanning Microscopies-Scanning, Vol. 17 (3), pp.
      175, 2006

[13] Pancove J. I., Optical Processes in Semiconductors, Prentice Hall, 1971

[14] Rajeev Bhatt S., Kar K., Bartwal S and Wadhawan V. K., Solid State
      Communications, Vol. 127, pp. 457, 2003. https://doi.org/10.1016/S0038-
      1098(03)00450-2

[15] Saravanan R., Private communication

[16] Sczancoski J. C., Bomio M. D. R., Cavalcante L. S., Joya M. R., Pizani P. S.,
      Varela J. A., Longo E., Siu Li M and Andrés J. A., J. Phys. Chem. C, Vol. 14, pp.
      113, 2009

[17] Scherrer P., Determination of the Size and Internal Structure of Colloidal Particles
      Using X-rays, Mathematisch-Physikalische Klasse, 1918

[18] Wood D. L and Tauc J., Phys. Rev., Vol. B5, pp. 3144, 1972.
      https://doi.org/10.1103/PhysRevB.5.3144

[19] Zhang Y., Holzwarth N. A. W and Williams R. T., Phys. Rev. B, Vol. 57, pp.
      12738, 199

Chapter 4

Analysis of Electron Density Distribution of Non-Linear Optical Materials

Abstract

Chapter IV deals with the electron density distribution analysis of non-linear optical materials such as $PbMoO_4$, $LiNbO_3$, $Ce:Gd_3Ga_5O_{12}$, $CaCO_3$, $Yb:CaF_2$, and Al_2O_3, $Cr:Al_2O_3$, $V:Al_2O_3$. The results of the electron density distribution studies are presented in the form of three, two dimensional electron density maps and one dimensional profile. This chapter provides quantitative and qualitative analysis of the bonding behaviour of the chosen non-linear optical materials.

Keywords

Electron Density, Optical Materials, Maximum Entropy Method, Mid Bond Electron Density, Charge Ordering

Contents

4.1 Introduction

The study of distribution of electrons around the atoms and their bonds in the bonding direction is an important component in materials characterization. The electronic structure of the system plays an important role in the optical and magnetic properties of the material. Ultimately, no study can manifest the legitimate picture, because no two experimental data are identical. This problem turns out to be intensified when the model used for the appraisement of electron densities is not completely applicable. The electron density $\rho(r)$ is defined as the number of electrons per unit volume.

An accurate electron density distribution is of utmost importance to understand the nature of bonds precisely. The very important statistical approach to deal with various crystallographic problems based on constrained entropy maximization, which constructs the electron density distribution is maximum entropy method (MEM) introduced by Collins (1982). Fourier synthesis of electron densities can also be of useful in picturizing the bonding between two atoms, but, it suffers from the major disadvantage of series termination error and negative electron densities that prevent the clear understanding of the fine nature of bonding between atoms, the factor which has been intended to be analyzed. The exact electron density distribution would be obtained if all the structure factors were known without ambiguities. An important issue in charge density analysis has always been the accuracy of the measured data. A high quality data set with a limited number of reflections can give reasonable MEM (Collins, 1982) electron density results. MEM (Collins, 1982) is a model free approach in contrast to structure refinements in which the positions of spherical atoms are determined. The principle of MEM (Collins, 1982) is to obtain an electron density distribution, which is consistent with the observed structure factors and to leave the uncertainties maximum. Also, no phase information is indispensable for MEM (Collins, 1982) calculations. Currently, MEM (Collins, 1982) formalism is being used as an effective tool for the visualization of bonding nature and the distribution of electrons in the bonding region with more accuracy in crystalline materials. Numerous research groups (Saravanan et al., 2003; Sakata et al., 1990; Yamamoto et al., 1996; Gilmore, 1996; Saka and Kato, 1986; Yamamura et al., 1998) have achieved precise results with experimental data sets using the maximum entropy method (Collins, 1982).

The study of the charge density and its distribution inside the unit cell plays a vital role for the better understanding of the chemical/physical properties. This visualization ability of high resolution charge density is useful to construct structure models of the materials. In this work, high resolution charge density distribution analysis was done using the maximum entropy method (MEM) (Collins, 1982). The observed structure factors extracted from the Rietveld (1969) refinement method were used for the MEM (Collins,

1982) procedure to obtain the charge density distribution in the unit cell. The Lagrange parameter is suitably chosen so that the convergence criterion C=1 is reached after minimum number of iterations. In our present study, the MEM (Collins, 1982) refinements were carried out by dividing the unit cell into suitable pixels along a, b and c-axes. The uniform prior density was used in all the cases by dividing the total number of electrons by the volume of the unit cell. The software PRIMA (PRactice of Interactive MEM Analysis) (Izumi and Dilanian, 2002) was used for the MEM computations. The results of the MEM refinement are visualized through the visualization software program VESTA (Visualization for Electronic and STructural Analysis) (Momma and Izumi, 2006). This program has provisions to study the electron density in different planes both qualitatively and quantitatively as the three-dimensional (3D), two-dimensional (2D) electron density distributions and one-dimensional (1D) profiles inside the unit cell of the crystalline material (Bricogne and Gilmore, 1990).

Hence, chapter IV is devoted for the analysis (results and discussion) of charge density distribution using the maximum entropy method (Collins, 1982) for the chosen non-linear optical materials. In the present work, the electron density distribution studies have been done for the following non-linear optical materials $PbMoO_4$, $LiNbO_3$, $Ce:Gd_3Ga_5O_{12}$, $CaCO_3$, $Yb:CaF_2$ and Al_2O_3, $Cr:Al_2O_3$ & $V:Al_2O_3$..

4.2 Electron density analysis of $PbMoO_4$

A high resolution pictorial analysis will give better insight of the study of the electron density distribution and bonding nature of the non-linear optical materials. The electron density analysis and the chemical bonding between the atoms has been elucidated for the non-linear optical material $PbMoO_4$, using the versatile technique MEM (Collins, 1982). The software program VESTA (Momma and Izumi, 2006) has been used for the visualization of three dimensional, two dimensional electron density distributions and 1D profile. The MEM refinement parameters of $PbMoO_4$ are given in table 4.2.1.

The unit cell of $PbMoO_4$ containing the constituent atoms Pb, Mo and O are shown in figure 4.2.1. The iso-surface levels are suppressed for the better view. The three-dimensional electron density distribution imposed on the structure in the form of iso-surface (iso-surface level: $1.0 \ e/\mathring{A}^3$) in the unit cell of $PbMoO_4$ was shown in figure 4.2.2. The unit cell of $PbMoO_4$ contains MoO_4 tetrahedra consisting of Mo atom at the centre and four oxygen atoms at the tetrahedral positions and PbO_8 octahedra consisting of Pb atoms in an octahedral arrangement with eight oxygen atoms surrounding Pb. The combination of these tetrahedral and octahedral arrangements decides the various properties of $PbMoO_4$. All the PbO_8 octahedra and MoO_4 tetrahedra in the unit cell are shown in figure 4.2.3. It is expected that the Pb-O bonding is of mixed covalent and ionic character and the Mo-O bonds are with covalent character (Sczancoski et al., 2009). The three-dimensional MEM electron density distribution containing all the PbO_8 octahedra in the unit cell of $PbMoO_4$ is shown in figure 4.2.4. Three-dimensional MEM electron density distribution containing the position of single PbO_8 octahedron in the unit cell of $PbMoO_4$ is shown in figures 4.2.5. All the eight oxygen atoms attached with the Pb atoms are visible in the octahedron. Different views of PbO_8 octahedra with iso-surface (iso-surface level: 1.3 $e/Å^3$) are shown in figures 4.2.6. The interaction of oxygen atoms can easily be noticed. PbO_8 octahedra surrounded by the electron clouds are presented in figure 4.2.6. The running wave like features in figure 4.2.6 is the interaction between Pb and Mo. These are heavy atoms and hence they extend themselves and interact with each other and also with oxygen atoms.

Table 4.2.1 Parameters from MEM refinement of $PbMoO_4$.

Parameter	Value
Number of cycles	673
Number of electrons/unit cell (F_{000})	624
Lagrange parameter (λ)	0.0836
R_{MEM} (%)	1.3922
$_wR_{MEM}$ (%)	1.8389

R_{MEM} - Reliability index from MEM refinement

$_wR_{MEM}$ - Weighted reliability index from MEM refinement

The three-dimensional MEM electron density distribution containing all the tetrahedra structures in the unit cell is shown in figure 4.2.7. Three-dimensional MEM electron density distribution containing the position of single MoO_4 tetrahedron in the unit cell of $PbMoO_4$ is shown in figure 4.2.8. Figure 4.2.9 shows the MoO_4 tetrahedra with a plane separating four oxygen atoms passing through the Mo atom situated at the centre of tetrahedron. The Pb atom lies at the edges of that plane. This plane is lying at a distance of 6.065 Å (half the unit cell along z-axis) from the origin. The two-dimensional electron density corresponding to this plane (Contour interval: 0 to 1 step size of 0.08 $e/Å^3$) is shown in figure 4.2.10. The heavy Mo and Pb atoms seem to mask the electron density of O atoms which lie in planes above and below the Mo atom at the centre of the tetrahedron. Different views of the same in the unit cell are shown within the boundary line of unit cell in figure 4.2.11. This (003) plane is lying at a distance of 2 Å from the origin. The two-dimensional electron density distribution on this plane (Contour interval:

0 to 5 step size of 0.48 e/Å³) is shown in figures 4.2.12 and 4.2.13. The presence and interaction of oxygen with Mo atom is visible through the contour lines, for e.g., at places near the elliptical voids in the figure 4.2.13.

Another view of the plane passing through the oxygen atoms, but in a direction perpendicular to it is shown in figure 4.2.14 and the corresponding 2D electron density distribution on that plane (Contour interval: 0 to 5 step size of 0.5 e/Å³) is shown in figure 4.2.15. The tetrahedron oxygen atoms are indicated using the sticks. The Mo atom is at the crossing of the line segments. The oxygen atoms are at the tips of the dark concentrated contour lines. The concentrated dark contour lines are due to Pb atoms. Figure 4.2.16 shows the (200) plane passing through the Mo and Pb atoms at a distance of 2.723 Å from the origin and the corresponding two-dimensional electron density distribution on the (200) plane is shown in figure 4.2.17 (Contour interval: 0 to 1 step size of 0.08 e/Å³). The tetrahedral positions of oxygen atoms can be inferred as shown in the figure 4.2.17. Figure 4.2.18 shows the (110) plane in the unit cell of PbMoO₄ and the corresponding two-dimensional electron density distribution on this (110) plane (Contour interval: 0 to 2 step size of 0.25 e/Å³) is shown in figure 4.2.19.

Figure 4.2.2. 3D electron density distribution imposed on the structure in the form of iso-surface in the unit cell of $PbMoO_4$ (iso-surface level: $1.0 \ e/Å^3$).

Pb

Mo

O

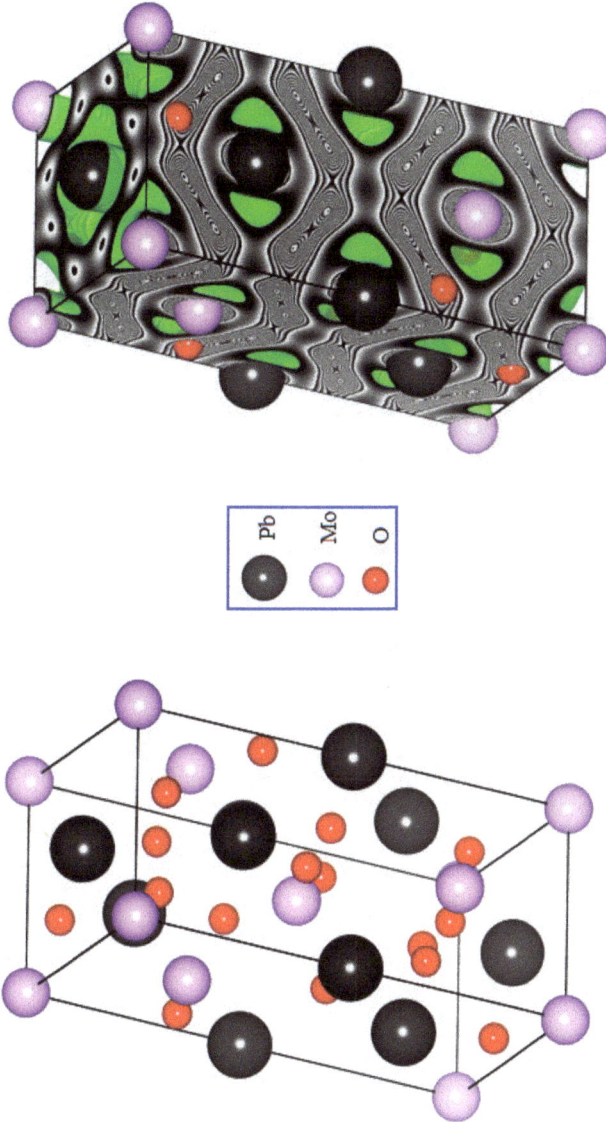

Figure 4.2.1 The unit cell of $PbMoO_4$.

111

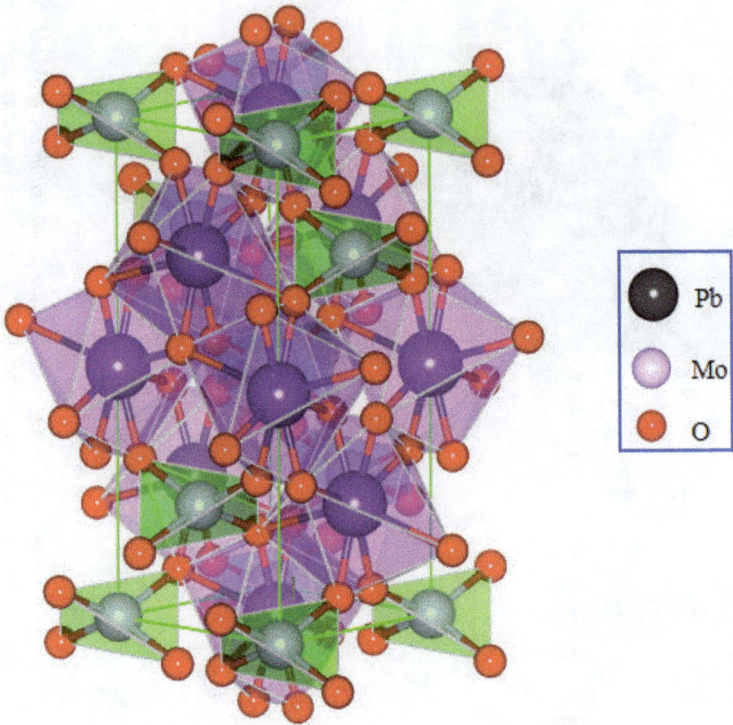

Figure 4.2.3 3D MEM electron density distribution containing all the PbO$_8$ octahedra and MoO$_4$ tetrahedra in the unit cell of PbMoO$_4$.

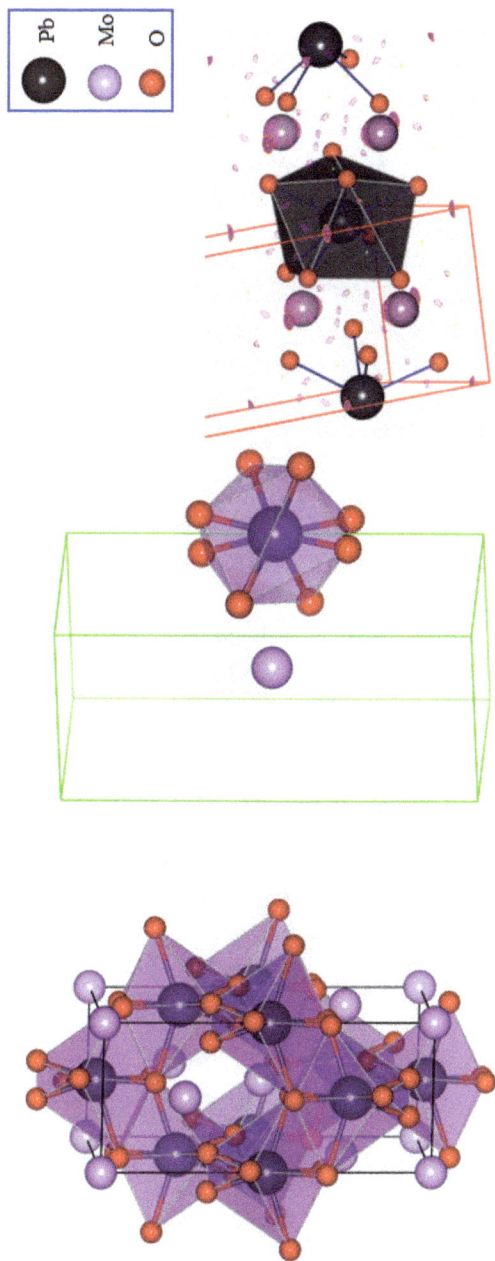

Figure 4.2.4 3D view containing all the PbO$_8$ octahedra in the unit cell of PbMoO$_4$.

Figure 4.2.5 3D MEM electron density distribution containing single PbO$_8$ octahedron in the unit cell of PbMoO$_4$.

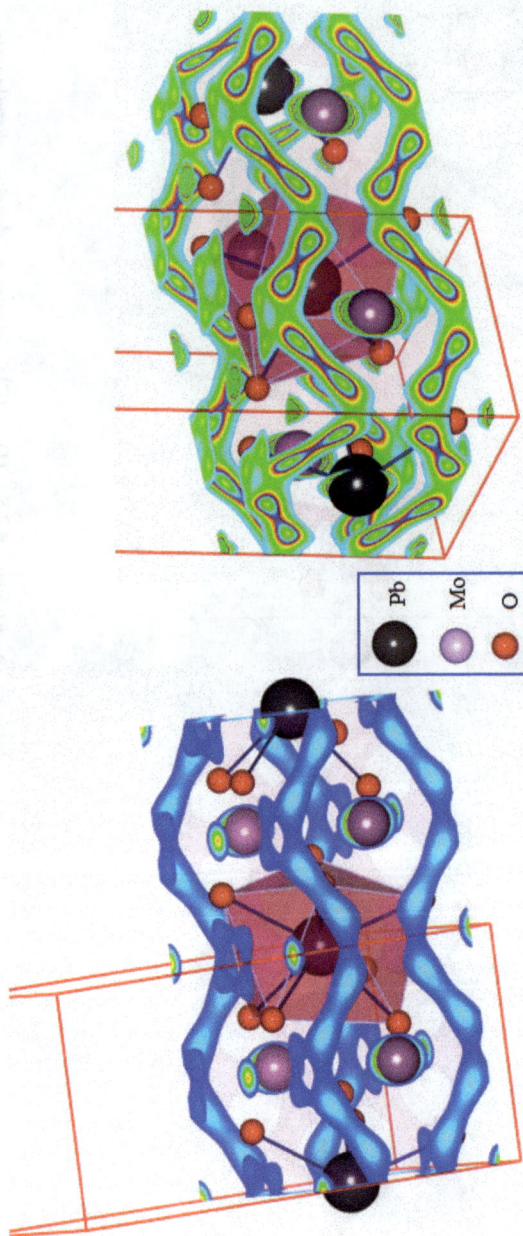

Figure 4.2.6 3D MEM electron density distribution containing single PbO$_8$ octahedron with iso-surface surrounded by the electron clouds (iso-surface level: 1.3 e/Å^3).

Figure 4.2.8 3D MEM electron density distribution containing single MoO$_4$ tetrahedron in the unit cell of PbMoO$_4$.

Pb
Mo
O

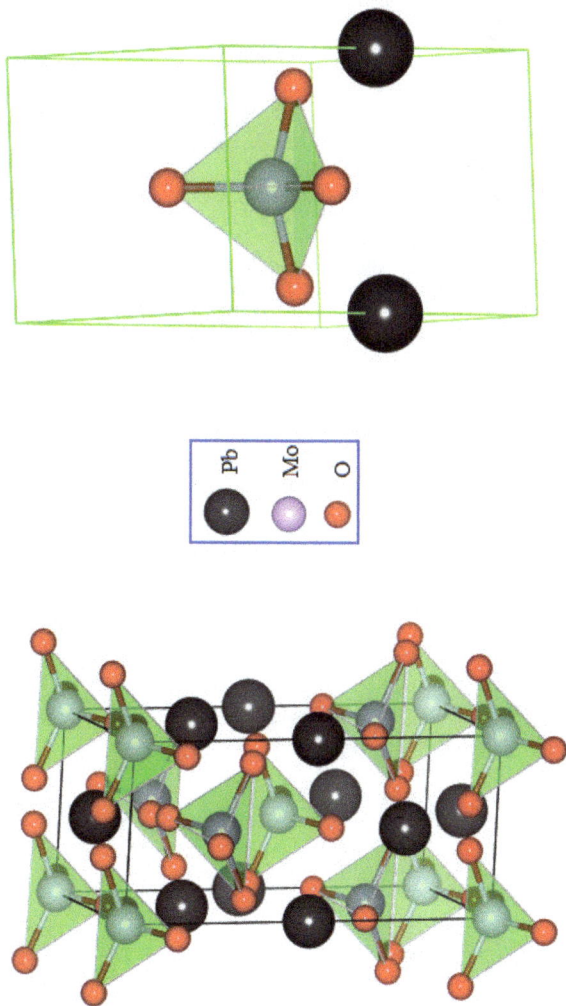

Figure 4.2.7 3D MEM electron density distribution containing all the MoO$_4$ tetrahedra in the unit cell of PbMoO$_4$.

Figure 4.2.10 2D electron density on the plane separating four oxygen atoms (Contour interval: 0 to 1 step size of 0.08 e/A³).

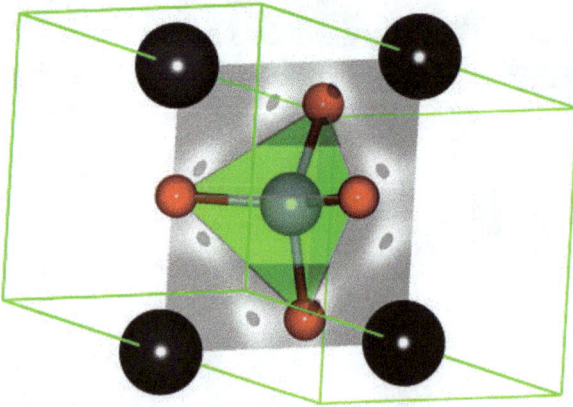

Figure 4.2.9 3D view of MoO₄ tetrahedron with a plane separating four oxygen atoms in the unit cell of PbMoO₄.

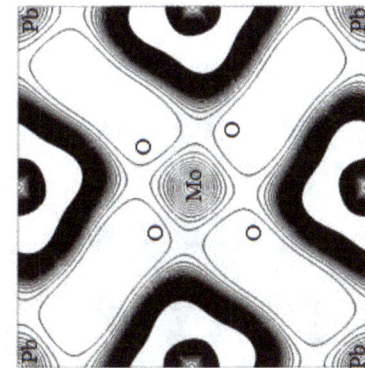

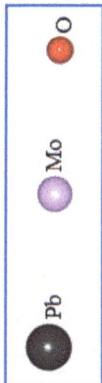

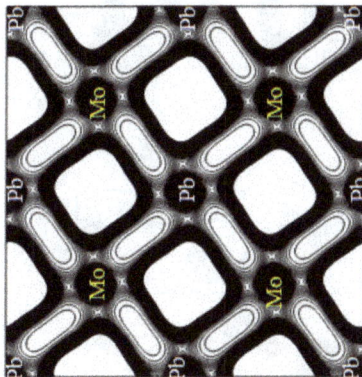

Figure 4.2.11 3D view of the unit cell shown within the boundary line.

Figure 4.2.12 2D electron density distribution on (003) plane (Contour interval: 0 to 5 step size of 0.48 e/Å³).

Figure 4.2.13 2D electron density distribution on (003) plane in the unit cell (Contour interval: 0 to 5 step size of 0.48 e/Å³).

117

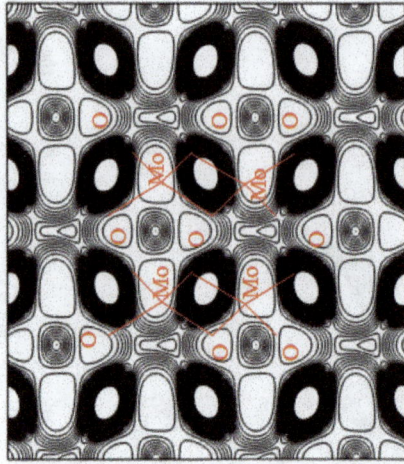

Figure 4.2.15 2D electron density distribution on the plane passing through oxygen atoms in the perpendicular direction (Contour interval: 0 to 5 step size of 0.5 e/Å³).

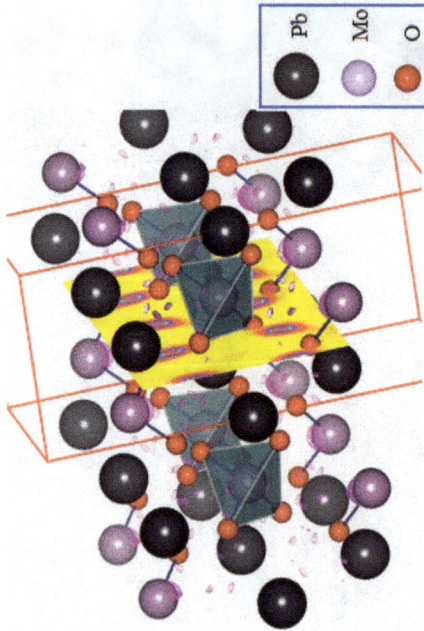

Pb	●
Mo	●
O	○

Figure 4.2.14 3D view of the plane passing through oxygen atoms in the perpendicular direction.

Figure 4.2.17 2D electron density distribution on the plane passing through the Mo and Pb atoms (Contour interval: 0 to 1 step size of 0.08 e/Å³).

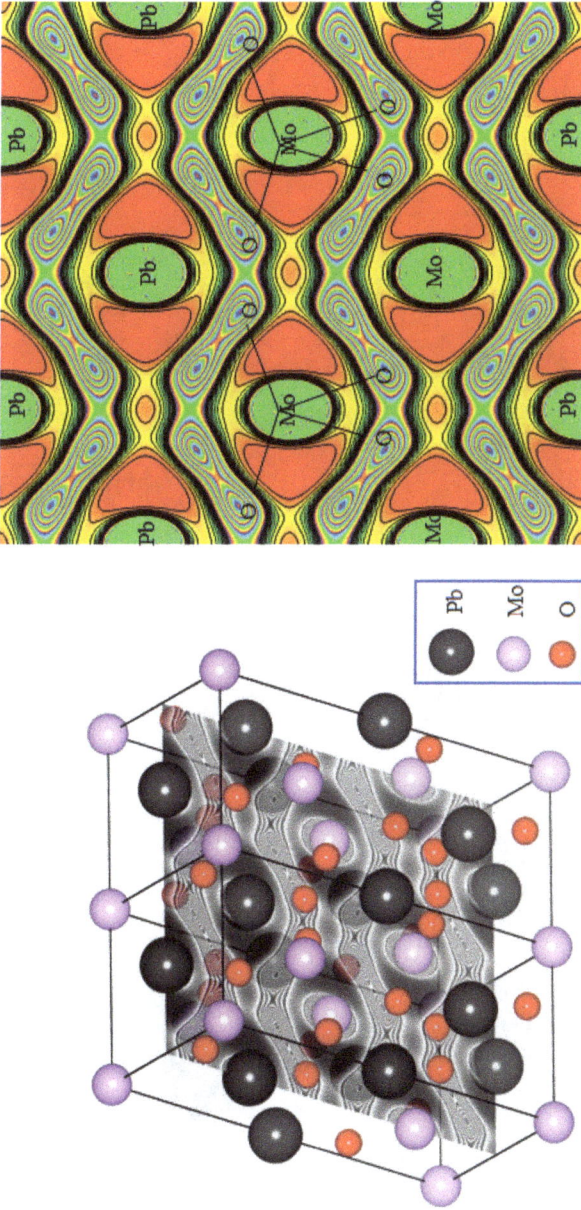

Figure 4.2.16 Plane passing through the Mo and Pb atoms in 3D.

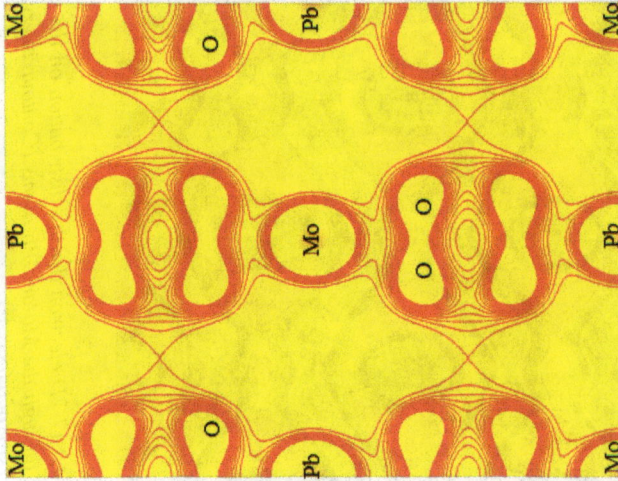

Figure 4.2.19 2D electron density distribution on (110) plane in the unit cell of PbMoO₄ (Contour interval: 0 to 2 step size of 0.25 e/Å³).

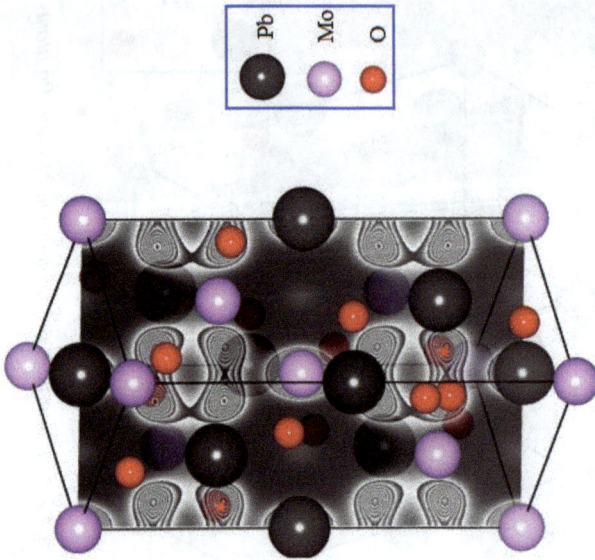

Figure 4.2.18 3D view of the unit cell of PbMoO₄ with (110) plane.

The numerical values of the MEM electron densities from one-dimensional electron density profiles along different bonding directions in the unit cell of $PbMoO_4$ are given in table 4.2.2. The one-dimensional MEM electron density profiles of PbMoO4 along different directions have been presented in figures 4.2.20 - 4.2.25. In all the 1D electron density profiles of $PbMoO_4$, the first atom has been considered to be in the origin.

Table 4.2.2 MEM electron densities of $PbMoO_4$ from 1D electron density analysis.

Direction	Position (Å)	Electron density (e/Å³)	Comment
Mo-Mo1	1.2323	0.1986	
	2.0855	1.5425	ED due to interaction of Mo-Mo
	3.0335	0.1251	
	3.9814	1.5425	ED due to interaction of Mo-Mo
	4.8346	0.1986	
Mo-Mo2	1.4165	0.0645	
	2.0382	3.1643	ED due to interaction of Mo-Mo
Mo-Mo3	1.1914	0.0042	
	2.1700	0.2139	ED due to interaction of Mo-Mo
	2.7223	0.1251	
	3.2763	0.2139	ED due to interaction of Mo-Mo
Mo-O1	1.1647	0.0103	
Mo-O2	1.2420	0.1951	
	2.1113	2.5727	ED due to interaction of Mo-O
	3.0739	0.2091	
	3.9743	3.8161	ED due to interaction of Mo-O
Mo-O3	1.3694	0.0091	
	1.7197	0.0116	ED due to nearby Pb atom
	2.3885	0.0047	
	3.7898	5.2069	High due to interaction of Pb atom
Mo-Pb1	1.1734	0.0055	
	1.9256	0.0901	Interaction of O atom involved
	2.6777	0.0055	
Mo-Pb2	1.2324	0.1986	
	2.0855	1.5425	ED due to interaction of Mo-Pb
	3.0335	0.1251	
	3.9815	1.5425	ED due to interaction of Mo-Pb
	4.8346	0.1986	
Pb-Pb1	1.1465	0.0645	
	2.0382	3.1643	ED due to interaction of O atoms from all sides
	2.9299	0.0645	
Pb-Pb2	1.2351	0.0649	
	2.0772	0.3290	Interaction of O atom involved

	3.5930	0.0901	Interaction of Mo atom involved
	5.1089	0.3290	Interaction of O atom involved
Pb-O1	1.1456	0.0141	
	2.4035	2.7907	ED due to interaction of Mo-O
	2.8753	3.8161	
Pb-O2	1.1999	0.1801	
	2.1175	4.3209	ED due to interaction of O atoms from all sides
O-O1	1.3704	3.8161	Interaction of O atom involved

ED - Electron density

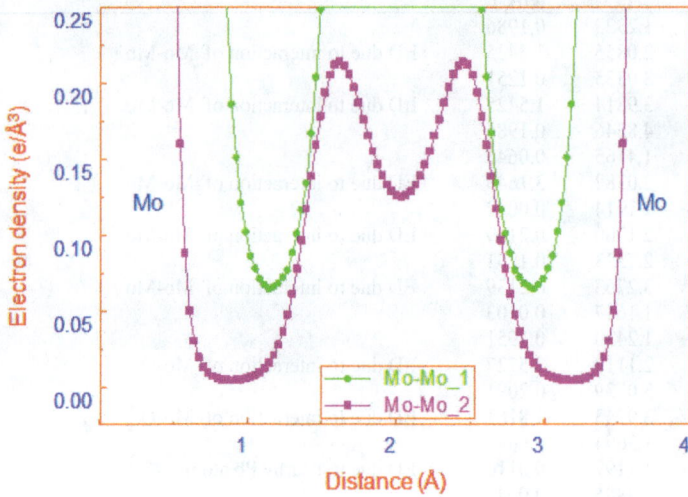

Figure 4.2.20 1D electron density profiles of Mo-Mo atoms in the unit cell of PbMoO₄.

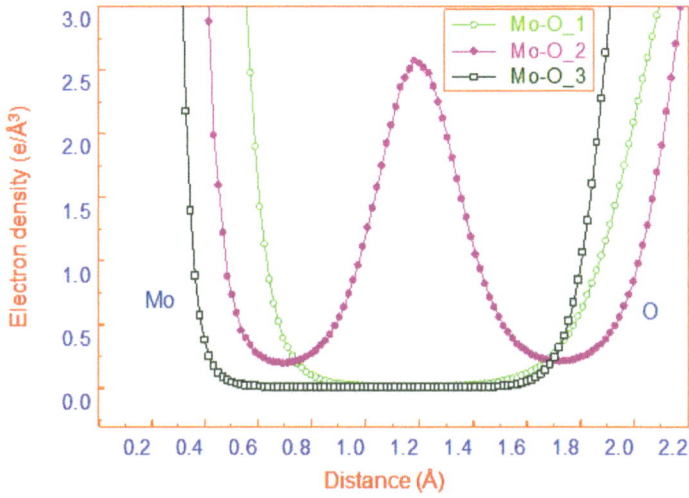

Figure 4.2.21 1D electron density profiles of Mo-O atoms in the unit cell of PbMoO$_4$.

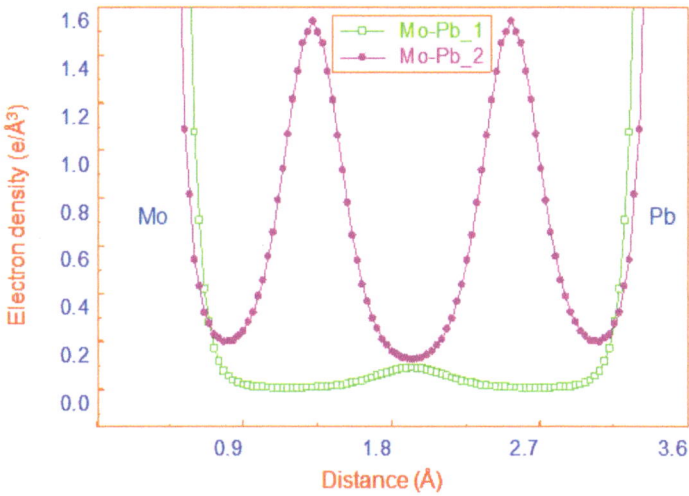

Figure 4.2.22 1D electron density profiles of Mo-Pb atoms in the unit cell of PbMoO$_4$.

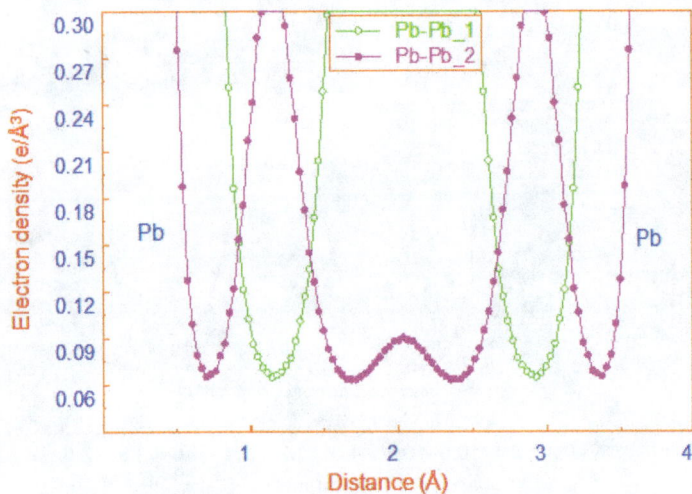

Figure 4.2.23 1D electron density profiles Pb-Pb atoms in the unit cell of PbMoO₄.

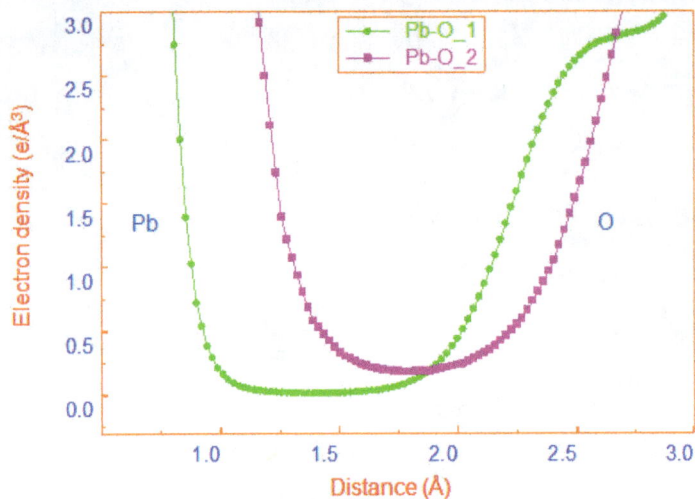

Figure 4.2.24 1D electron density profiles of Pb-O atoms in the unit cell of PbMoO₄.

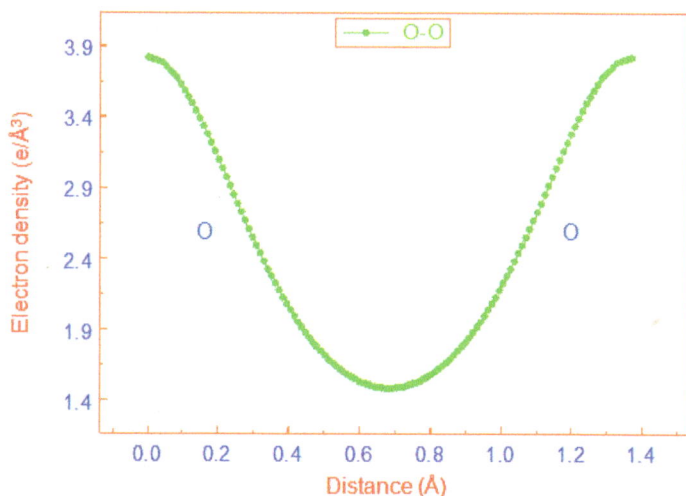

Figure 4.2.25 1D electron density profile of O-O atoms in the unit cell of PbMoO₄.

4.3 Electron density analysis of LiNbO₃

The maximum entropy method (MEM) (Collins, 1982) is one of the appropriate methods in which the concept of entropy is introduced to handle the uncertainty properly. It has been suggested that MEM (Collins, 1982) would be a suitable method for examining electron densities in the outer atomic region, for example, bonding region.

The MEM refinements for LiNbO₃were carried out by dividing the unit cell into 128x128x128 pixels. The MEM (Collins, 1982) parameters of LiNbO₃ are given in table 4.3.1. The three-dimensional electron density imposed on the structure of LiNbO₃ in the unit cell is shown in figure 4.3.1. The three-dimensional electron density distribution imposed on the structure in the form of iso-surface (iso-surface level: 0.6 e/Å^3) in the unit cell of LiNbO₃ has been represented in figure 4.3.2. The (110) plane passing through the unit cell of LiNbO₃ is shown in figure 4.3.3. The two-dimensional electron density distribution corresponding to this (110) (Contour interval: 0 to 1.2 step size of 0.07 e/Å^3) plane has been given in figure 4.3.4. Figure 4.3.5 represents the (110) plane passing through the unit cell of LiNbO₃. The two-dimensional electron density distribution corresponding to this (110) plane (Contour interval: 0 to 1 step size of 0.08 e/Å^3) plane has been given in figures 4.3.6. Figure 4.3.7 represents (001) plane passing through the unit cell of $LiNbO_3$ lying at a distance of 13.8741 Å from the origin and the corresponding two-dimensional electron density distribution map has also been given in figure 4.3.8 (Contour interval: 0 to 3 step size of 0.25 $e/Å^3$). Figure 4.3.9 represents (001) plane lying at a distance of 13.2141 Å from the origin and the two-dimensional electron density distribution map corresponding to this plane has been given in figure 4.3.10 (Contour interval: 0 to 3 step size of 0.25 $e/Å^3$). The oxygen atoms attached to the Nb atom are not on the same plane. The bonding tendency between Nb and O atoms can easily be visualized in the two-dimensional picture which has been drawn on (001) plane. The weak interaction of Li atom with Nb and O atoms can be easily visualized through the contour lines in all the above planes.

Table 4.3.1 Parameters from MEM refinement of $LiNbO_3$.

Parameter	Value
Number of cycles	394
Lagrange parameter (λ)	0.2465
Number of electrons/unit cell (F_{000})	408
R_{MEM} (%)	0.034

R_{MEM} - Reliability index from MEM refinement

$_wR_{MEM}$ - Weighted reliability index from MEM refinement

Figure 4.3.2 3D electron density distribution imposed on the structure in the form of iso-surface in the unit cell of LiNbO$_3$ (iso-surface level: 0.6 e/Å^3).

Li	Nb	O
🟢	⚫	🟠

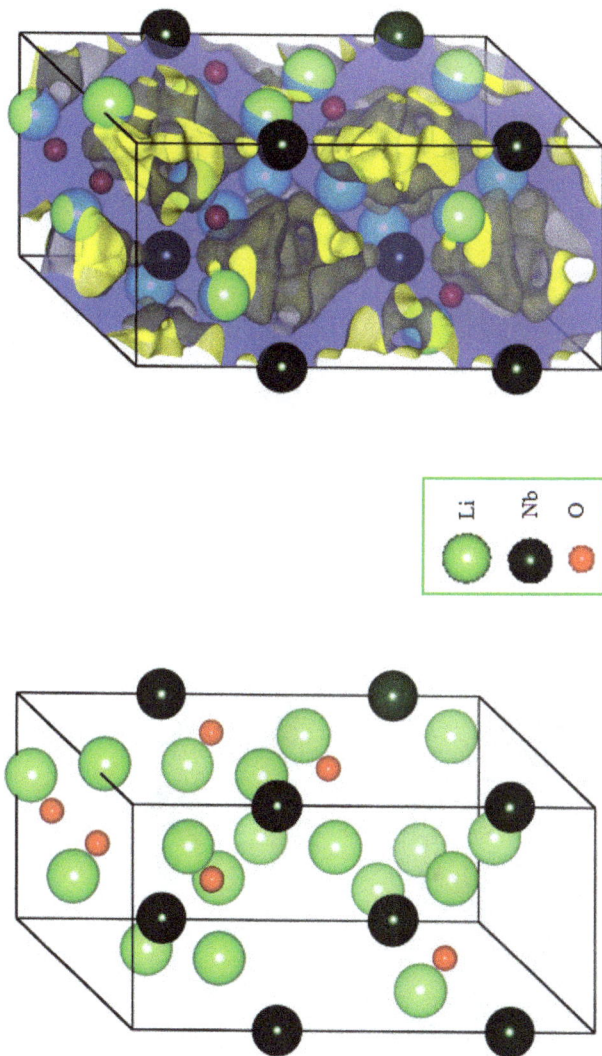

Figure 4.3.1 The unit cell of LiNbO$_3$.

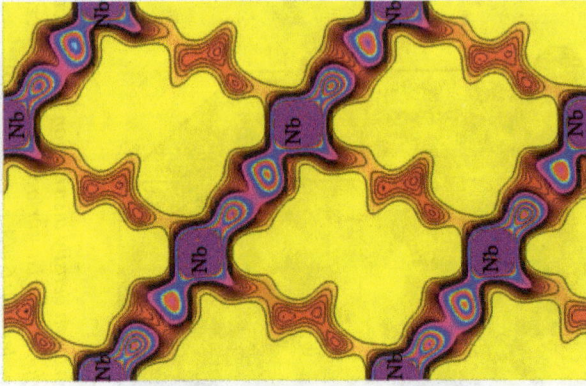

Figure 4.3.4 2D electron density distribution on (110) plane in the unit cell of LiNbO₃ (Contour interval: 0 to 1.2 step size of 0.07 e/Å³).

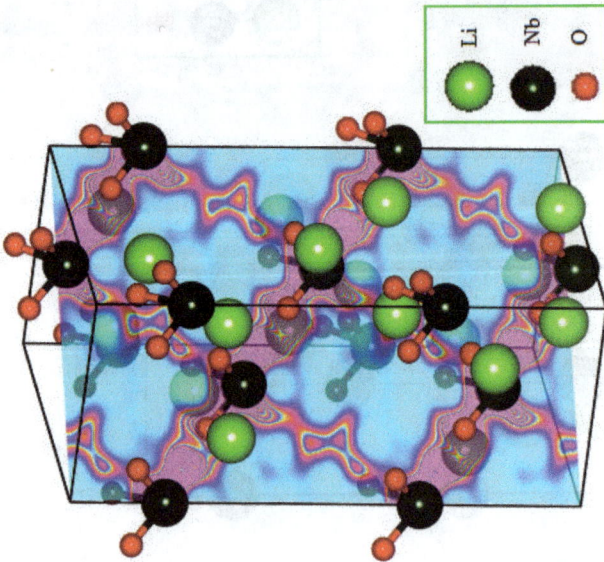

Figure 4.3.3 3D view of the (110) plane passing through the unit cell of LiNbO₃.

Li Nb O

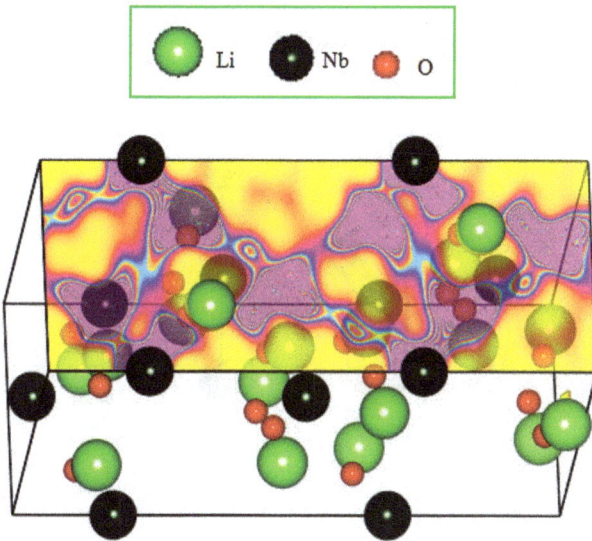

Figure 4.3.5 3D view of the (110) plane in the unit cell of LiNbO₃

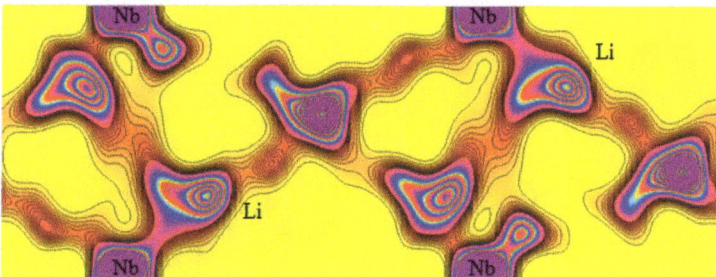

Figure 4.3.6 2D electron density distribution on (110) plane in the unit cell of LiNbO₃ (Contour interval: 0 to 1 step size of 0.08 e/Å³).

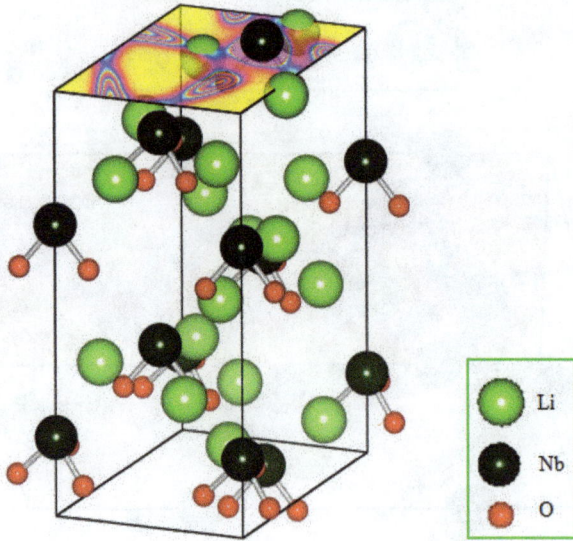

Figure 4.3.7 3D view of the (001) plane at a distance of 13.8741 Å in the unit cell of LiNbO₃.

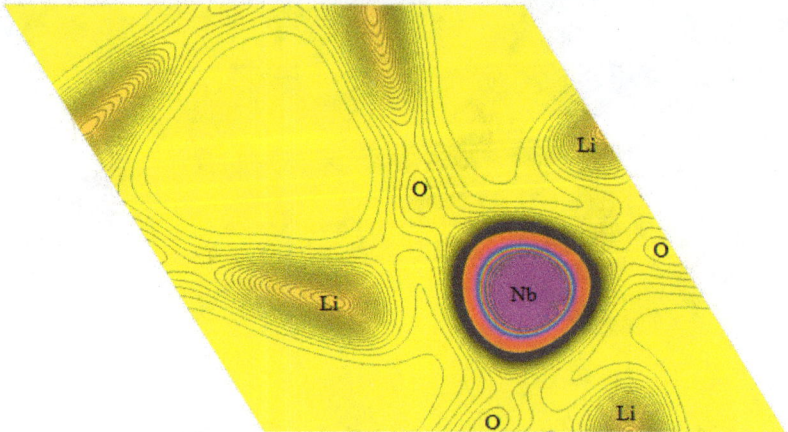

Figure 4.3.8 2D electron density distribution on (001) plane at a distance of 13.8741 Å in the unit cell of LiNbO₃ (Contour interval: 0 to 3 step size of 0.25 e/Å³).

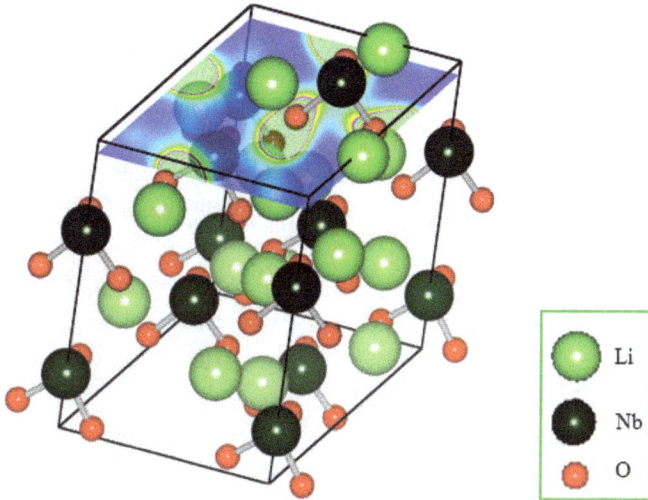

Figure 4.3.9 3D view of the (001) plane at a distance of 13.2141 Å in the unit cell of LiNbO₃.

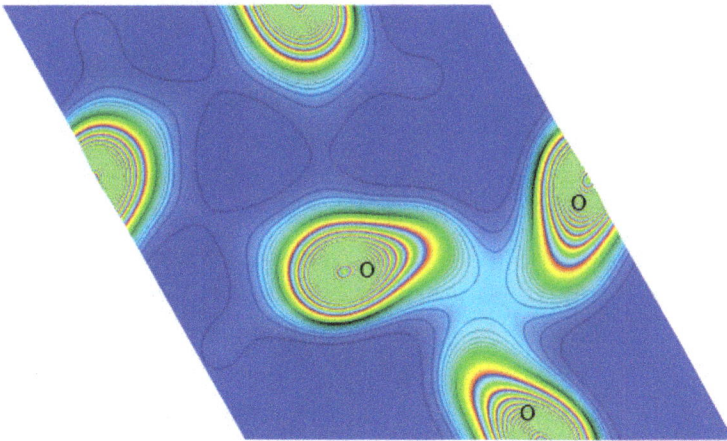

Figure 4.3.10 2D electron density distribution on (001) plane at a distance of 13.2141 Å in the unit cell of LiNbO₃ (Contour interval: 0 to 3 step size of 0.25 e/Å³).

Table 4.3.2 MEM bond densities of LiNbO$_3$ from 1D electron density profiles.

Direction	Position (Å)	Electron density (e/Å^3)	Comment
Li–Li	0.4295	0.1993	Max. ED along Li-Li bond
	1.7608	0.0107	Min. ED along Li-Li bond
	1.9111	0.0113	Peak due to Li-Li interaction
	2.7485	0.0817	Peak due to Li-Li interaction
Li–O	0.8927	0.0053	Min. ED between Li and O atoms
	1.9936	6.0650	Peak due to O atom
Nb–Li	0.0000	127.4353	Peak due to Nb atom
	0.9369	1.2987	Peak due to Li atom
Nb–Nb	0.0000	127.4353	Peak due to Nb atom
	1.1674	0.0979	Peak due to Nb-Nb interaction
	2.6568	0.3852	Peak due to Nb-Nb interaction
	3.8644	0.1178	Peak due to Nb-Nb interaction
	5.1525	127.4353	Peak due to Nb atom
Nb–O	0.0000	127.4353	Peak due to Nb atom
	1.9471	1.0127	Peak due to Nb-O interaction
	2.7488	0.2327	Peak due to Nb-O interaction
	3.3788	0.5703	Peak due to Nb-O interaction
	3.8369	0.3397	Peak due to Nb-O interaction
	4.5814	5.4400	Peak due to O atom
O–O	0.0000	6.0544	Peak due to O atom
	0.8447	0.4001	Peak due to O-O interaction
	0.9050	0.4085	Peak due to O-O interaction
	0.9653	0.4135	Peak due to O-O interaction
	1.8703	0.0729	Peak due to O-O interaction
	2.8356	2.2757	Peak due to O atom

ED - Electron density

The numerical values of the MEM electron densities from one-dimensional electron density profiles along different bonding directions in the unit cell of LiNbO$_3$ are given in table 4.3.2. The one-dimensional MEM electron density profiles of $LiNbO_3$ between various atoms have been presented in figures 4.3.11 – 4.3.16.

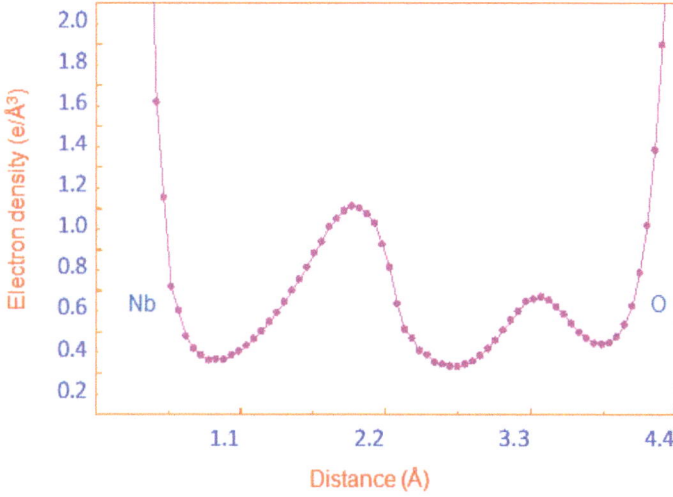

Figure 4.3.11 1D electron density profile of Nb-O atoms in the unit cell of $LiNbO_3$.

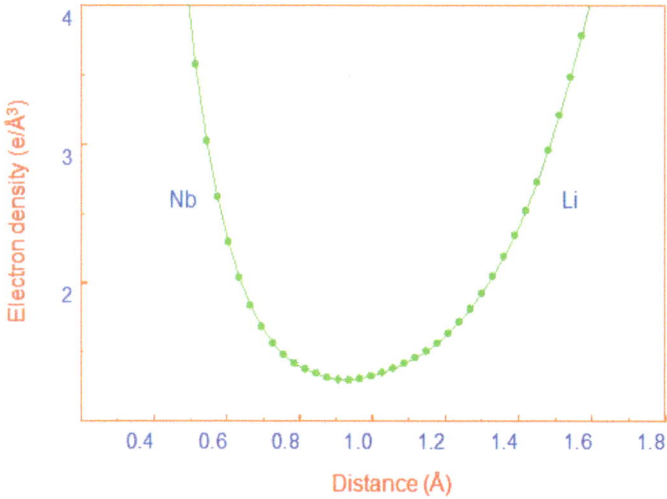

Figure 4.3.12 1D electron density profile of Nb-Li atoms in the unit cell of $LiNbO_3$.

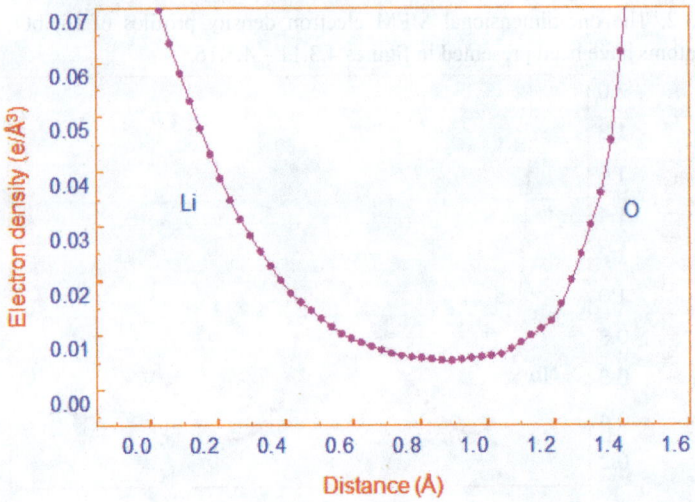

Figure 4.3.13 1D electron density profile of Li-O atoms in the unit cell of LiNbO₃.

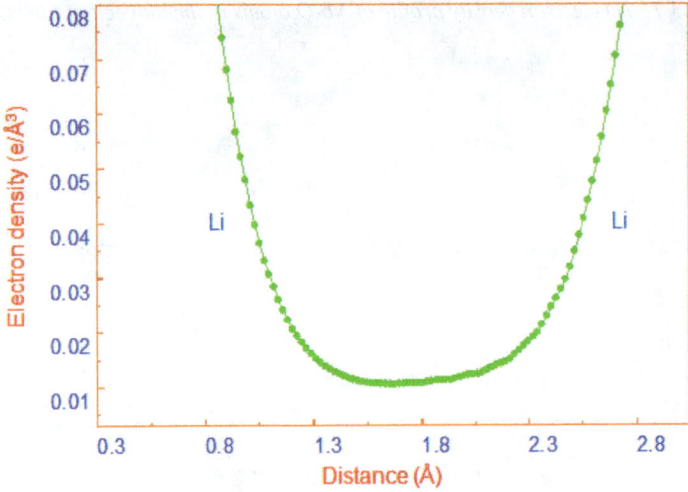

Figure 4.3.14 1D electron density profile of Li-Li atoms in the unit cell of LiNbO₃.

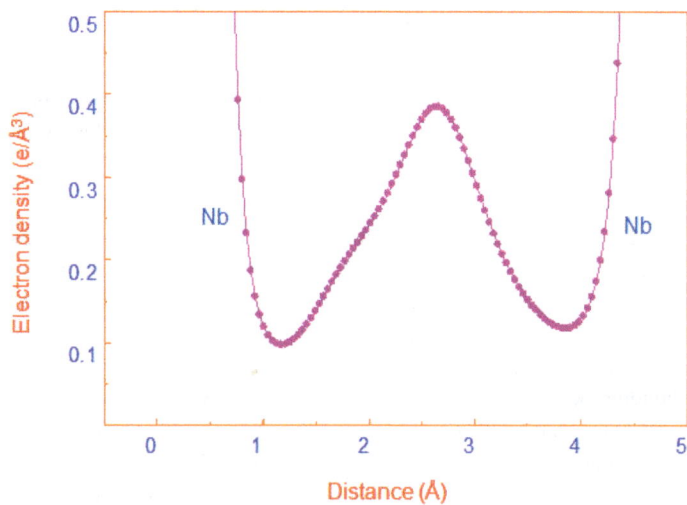

Figure 4.3.15 1D electron density profile of Nb-Nb atoms in the unit cell of LiNbO₃.

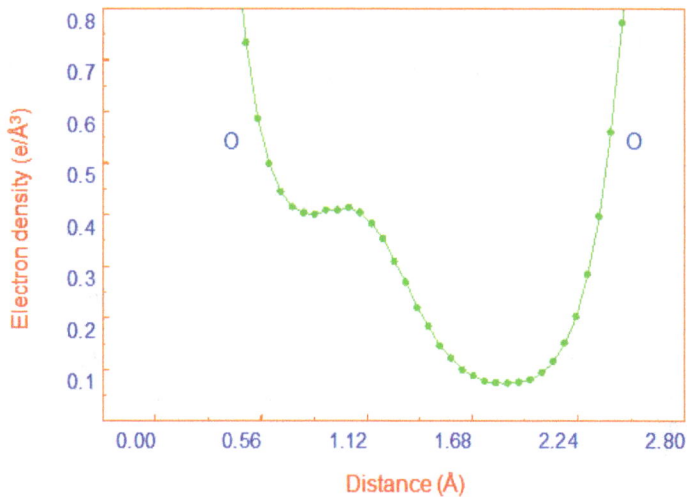

Figure 4.3.16 1D electron density profile of O-O atoms in the unit cell of LiNbO₃.

4.4 Electron density analysis of $Gd_{3-x}Ce_xGa_5O_{12}$

The maximum entropy method (MEM) (Collins, 1982) is an ingenious technique to study the electron density distribution in the unit cell. The advantage of this method is the clear visualization of bonding nature and the distribution of electrons in the bonding region with more accuracy. The obtained MEM parameters of $Gd_{3-x}Ce_xGa_5O_{12}$ (x = 0.5, 1 and 3) have been given in table 4.4.1.

Figure 4.4.1 represents the individual atoms imposed on the unit cell of $Gd_3Ga_5O_{12}$ and the iso-surface levels are suppressed for the better view. In figures 4.4.2 - 4.4.4, the regions surrounding each atom represent the charge density distributions of $Gd_{3-x}Ce_xGa_5O_{12}$ (x = 0.5, 1 and 3) considering the same iso-surface level of 1.5 $e/Å^3$ respectively. The chosen planes in this discussion were selected in order to reveal the nature of bonding between the pair of different atoms of $Gd_{3-x}Ce_xGa_5O_{12}$ (x = 0.5, 1 and 3). Figure 4.4.5 shows the (001) plane passing through the unit cell (half-the unit cell of $Gd_3Ga_5O_{12}$). Figures 4.4.6 - 4.4.8 show the two- dimensional electron density distributions of $Gd_{3-x}Ce_xGa_5O_{12}$ (x = 0.5, 1 and 3) on the (001) plane respectively. The contour level ranges from 0.0 to 6.0 $e/Å^3$ with the interval of 0.15 $e/Å^3$. The peak density of O atom increases with the increase of the trivalent Ce^{3+} atom and the bonding between Gd and O within the dodecahedron (GdO_8) must be covalent. The 3D view of the unit cell (half-the unit cell of $Gd_3Ga_5O_{12}$) with the (202) plane at a distance of 4.205 Å away from the origin and passing through the Ga(1) and O atoms of the cell is shown in figure 4.4.9. Figures 4.4.10 - 4.4.12 show the two-dimensional electron density distributions of $Gd_{3-x}Ce_xGa_5O_{12}$ (x = 0.5, 1 and 3) on the (202) plane respectively. The contour level ranges from 0.0 $e/Å^3$ to 3.0 $e/Å^3$ with the interval of 0.11 $e/Å^3$. Saddles with a little flattening at the mid bond positions in figures 4.4.10 - 4.4.12 show that bonding between Ga(1) and O is essentially covalent within $Ga(1)O_6$ octahedral arrangement. Figure 4.4.13 shows the (201) plane passing through all the constituent atoms within the unit cell (half-the unit cell of $Gd_3Ga_5O_{12}$) at a distance of 5.621 Å from the origin. Figures 4.4.14 - 4.4.16 show the two- dimensional electron density distributions of $Gd_{3-x}Ce_xGa_5O_{12}$ (x = 0.5, 1 and 3) on the (201) plane respectively. The contour level ranges from 0.0 $e/Å^3$ to 7.0 $e/Å^3$ with the interval of 0.27 $e/Å^3$. This variation of electron density distribution between different pairs of atoms can be clearly visualized through the contour lines of the two-dimensional electron density distributions.

Table 4.4.1 Parameters from MEM refinements of $Gd_{3-x}Ce_xGa_5O_{12}$.

Parameter	Composition of Ce		
	0.5%	1%	3%
Number of cycles	5929	3756	3691
Lagrange parameter (λ)	0.0086	0.0475	0.0469
R_{MEM} (%)	1.85	2.25	2.49
$_wR_{MEM}$ (%)	1.88	2.13	2.23

R_{MEM} - Reliability index from MEM refinement

$_wR_{MEM}$ - Weighted reliability index from MEM refinement

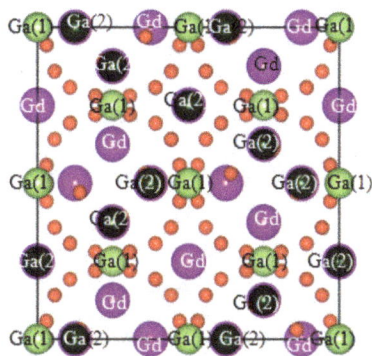

Figure 4.4.1 Unit cell of $Gd_3Ga_5O_{12}$.

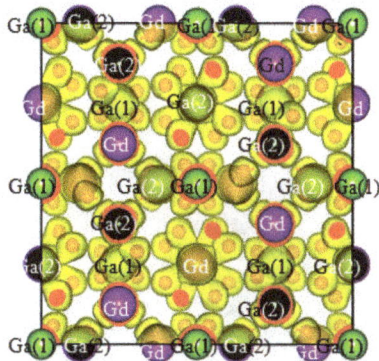

Figure 4.4.2 Unit cell of $Gd_{3-x}Ce_xGa_5O_{12}$ (x = 0.5) (iso-surface level: 1.5 $e/Å^3$).

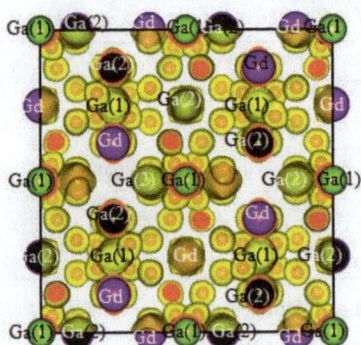

Figure 4.4.3 Unit cell of Gd$_{3-x}$Ce$_x$Ga$_5$O$_{12}$ (x = 1) (iso-surface level: 1.5 e/$Å^3$).

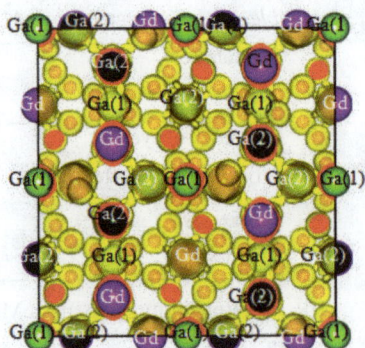

Figure 4.4.4 Unit cell of Gd$_{3-x}$Ce$_x$Ga$_5$O$_{12}$ (x = 3) (iso-surface level: 1.5 e/$Å^3$).

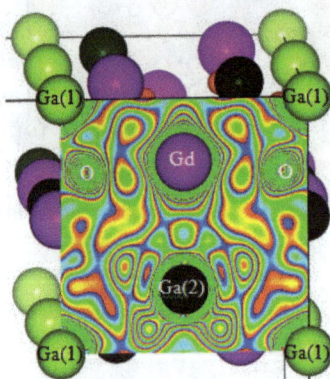

Figure 4.4.5 3D view of the (001) plane in the unit cell of Gd$_3$Ga$_5$O$_{12}$.

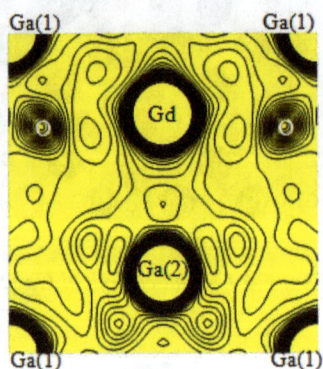

Figure 4.4.6 2D electron density distribution on the (001) plane in the unit cell of Gd$_{3-x}$Ce$_x$Ga$_5$O$_{12}$ (x = 0.5).

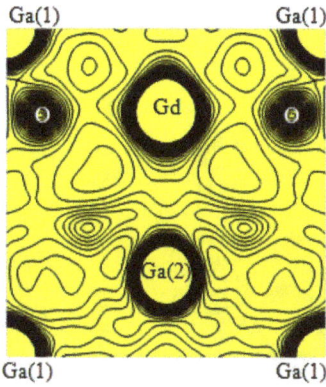

Figure 4.4.7 2D electron density distribution on the (001) plane in the unit cell of $Gd_{3-x}Ce_xGa_5O_{12}$ (x = 1).

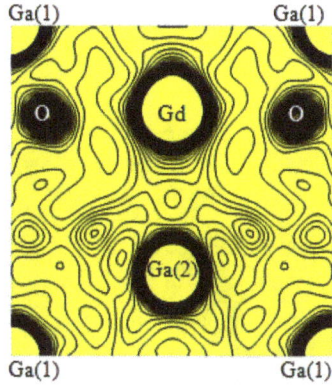

Figure 4.4.8 2D electron density distribution on the (001) plane in the unit cell of $Gd_{3-x}Ce_xGa_5O_{12}$ (x = 3).

(Contour interval: 0 to 6 step size of 0.15 e/$Å^3$)

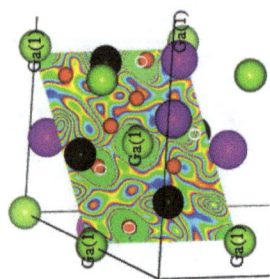

Figure 4.4.10 2D electron density distribution on the (202) plane in the unit cell of $Gd_{3-x}Ce_xGa_5O_{12}$ (x = 0.5).

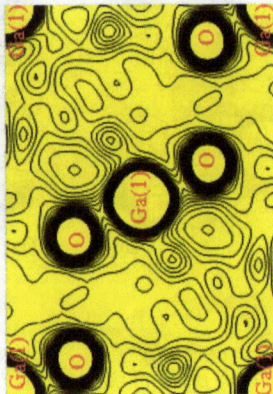

Figure 4.4.12 2D electron density distribution on the (202) plane in the unit cell of $Gd3-xCexGa5O12$ (x = 3).

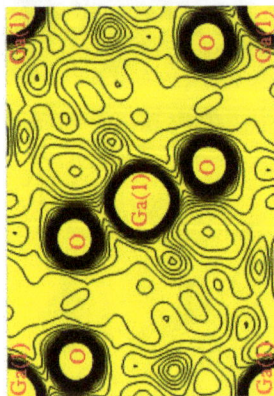

Figure 4.4.9 3D view of the (202) plane in the unit cell of $Gd_3Ga_5O_{12}$.

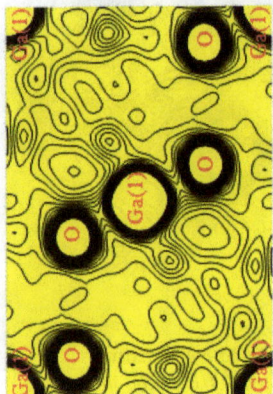

Figure 4.4.11 2D electron density distribution on the (202) plane in the unit cell of $Gd3-xCexGa5O12$ (x = 1).

(Contour interval: 0 to 3 step size of 0.11 e/$Å^3$)

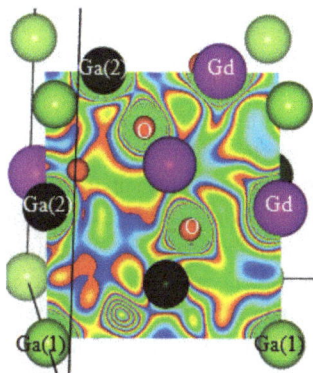

Figure 4.4.13 3D view of the (201) plane in the unit cell of Gd₃Ga₅O₁₂.

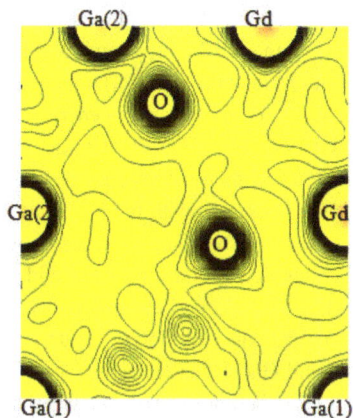

Figure 4.4.14 2D electron density distribution on the (201) plane in the unit cell of $Gd_{3-x}Ce_xGa_5O_{12}$ (x = 0.5).

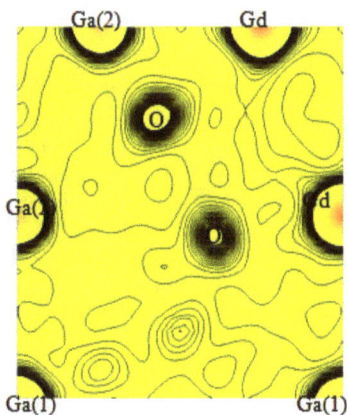

Figure 4.4.15 2D electron density distribution on the (201) plane in the unit cell of $Gd_{3-x}Ce_xGa_5O_{12}$ (x = 1).

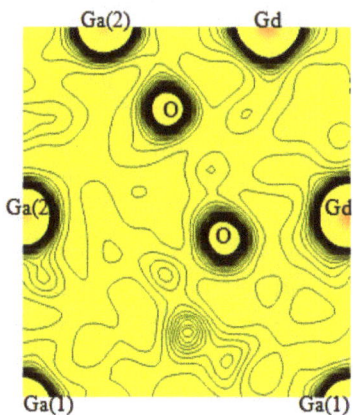

Figure 4.4.16 2D electron density distribution on the (201) plane in the unit cell of $Gd_{3-x}Ce_xGa_5O_{12}$ (x = 3).

(Contour interval: 0 to 7 step size of 0.27 e/$Å^3$)

One dimensional electron density profiles drawn between different pair of atoms are shown in figures 4.4.17 - 4.4.24. The mid-bond positions and the calculated numerical values of the MEM mid-bond densities between the pair of different atoms of $Gd_{3-x}Ce_xGa_5O_{12}$ (x = 0.5, 1 and 3) obtained from 1D electron density profiles are presented in table 4.4.2. These values in general, show that the mid-bond electron densities between Gd–d, Gd–Ga(2) and O–O decrease with the increase of the nominal concentration of Ce up to 1% and then increase for 3%. The mid-bond electron density between Gd–O and Ga(1)–O increases with the increase of nominal concentration of Ce up to 1% and then decreases for 3%. These values show typically the bonding to be covalent with the increasing electron density at the mid bond position with the increasing concentration of the Ce. The mid-bond electron density between Ga(2)–O decreases with the increase of the nominal concentration of Ce up to 3%. The mid-bond electron density between Gd–Ga(1) and Ga(1)–Ga(2) decreases with the increase of the nominal concentration of Ce up to 3%. The bond lengths between the atoms are also modified with Ce doping effect and it leads to increase in the volume of the unit cell.

Table 4.4.2 MEM mid-bond electron densities between different atoms of $Gd_{3-x}Ce_xGa_5O_{12}$ obtained from 1D electron density profiles.

Bonding atom pair	Composition of Ce					
	0.5		1		3	
	Position (Å)	Mid-bond electron density (e/Å³)	Position (Å)	Mid-bond electron density (e/Å³)	Position (Å)	Mid-bond electron density (e/Å³)
Gd–Gd	3.4724	0.6402	3.4724	0.5217	3.4724	0.5931
Gd–Ga(1)	1.5192	0.2742	1.7362	0.2825	1.6277	0.3865
Gd–Ga(2)	1.7470	0.5925	1.5529	0.5217	1.7470	0.5654
Gd–O	1.4727	0.2962	1.5758	0.6355	1.5307	0.5043
Ga(1)–O	1.0710	0.7808	0.9857	0.8868	1.0220	0.6504
Ga(2)–O	0.9732	0.8327	1.0606	0.7289	1.0156	0.5480
Ga(1)–Ga(2)	1.5130	0.3124	1.7291	0.3353	1.7291	0.4997
O–O	2.0702	1.2202	2.0702	0.4989	2.0702	1.0105

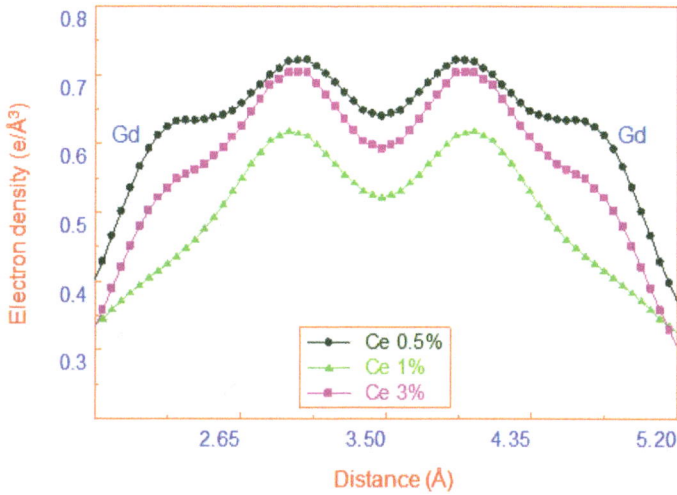

Figure 4.4.17 1D electron density profiles of Gd-Gd atoms in the unit cell of $Gd_{3-x}Ce_xGa_5O_{12}$.

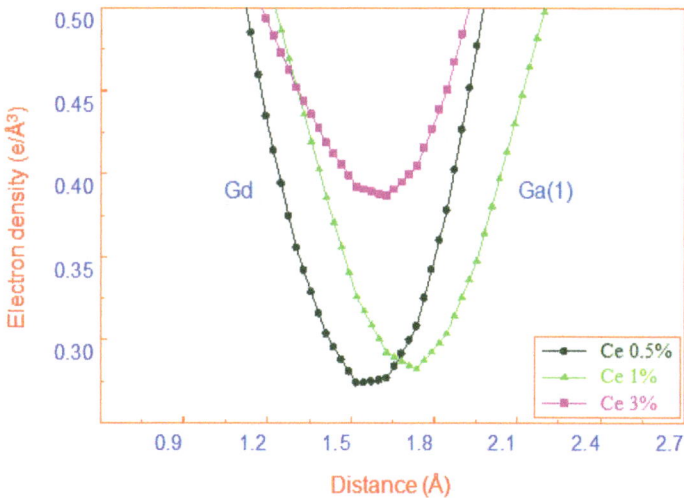

Figure 4.4.18 1D electron density profiles of Gd-Ga(1) atoms in the unit cell of $Gd_{3-x}Ce_xGa_5O_{12}$.

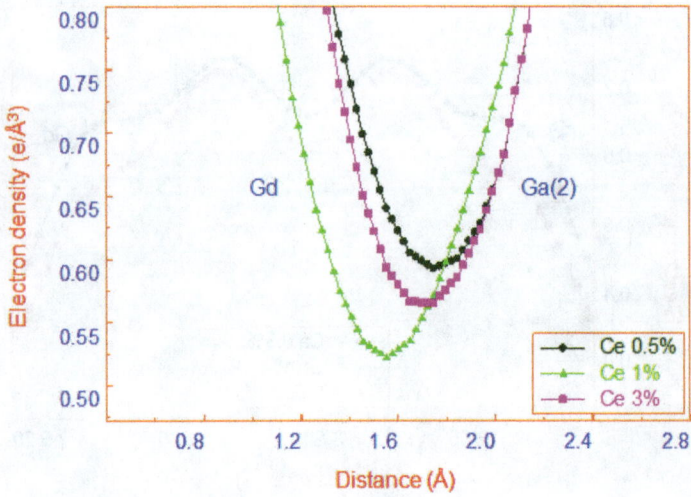

Figure 4.4.19 1D electron density profiles of Gd-Ga(2) atoms in the unit cell of $Gd_{3-x}Ce_xGa_5O_{12}$.

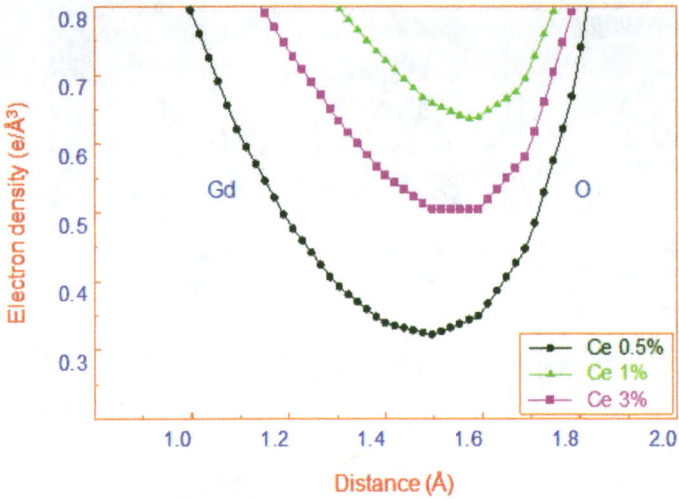

Figure 4.4.20 1D electron density profiles of Gd-O atoms in the unit cell of $Gd_{3-x}Ce_xGa_5O_{12}$.

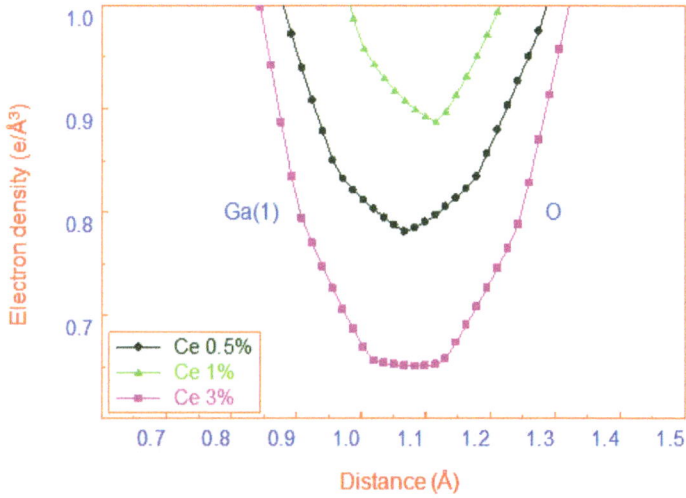

Figure 4.4.21 1D electron density profiles of Ga(1)-O atoms in the unit cell of $Gd_{3-x}Ce_xGa_5O_{12}$.

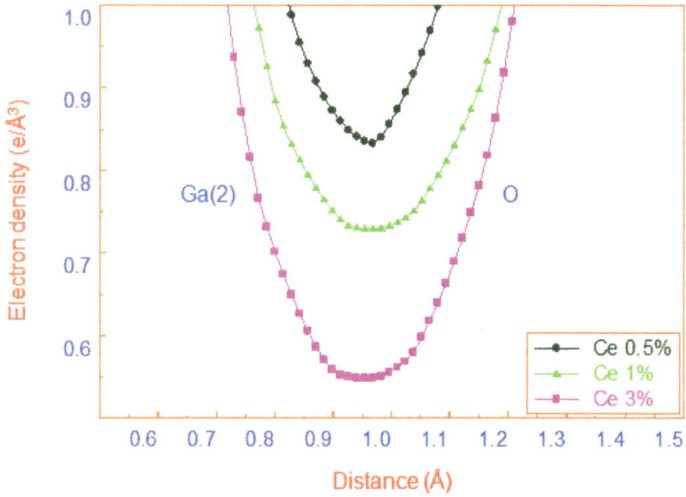

Figure 4.4.22 1D electron density profiles of Ga(2)-O atoms in the unit cell of $Gd_{3-x}Ce_xGa_5O_{12}$.

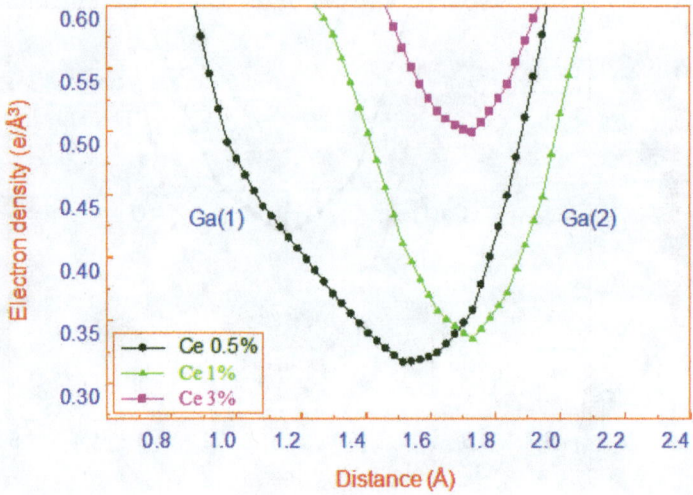

Figure 4.4.23 1D electron density profiles of Ga(1)-Ga(2) atoms in the unit cell of $Gd_{3-x}Ce_xGa_5O_{12}$.

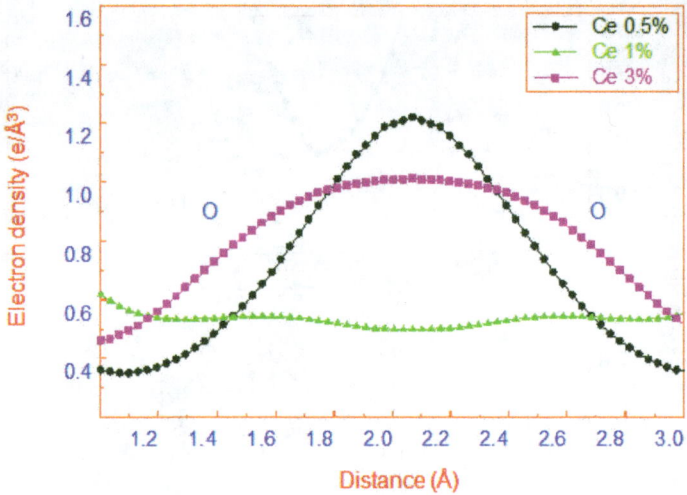

Figure 4.4.24 1D electron density profiles of O-O atoms in the unit cell of $Gd_{3-x}Ce_xGa_5O_{12}$.

4.5 Electron density analysis of CaCO₃

The electronic level charge density analysis and the chemical bonding between the atoms were elucidated for the thermoelectric material calcite using the versatile technique MEM (Collins, 1982). The uniform prior density was used by dividing the total number of electrons by the volume of the unit cell. The obtained MEM (Collins, 1982) parameters of calcite have been given in table 4.5.1.

Figure 4.5.1 represents the individual atoms imposed on the unit cell of calcite and the iso-surface level is suppressed for better view. Three dimensional electron density distributions on the unit cell of calcite have been represented in figure 4.5.2 considering the iso-surface level of 0.15 e/Å^3. The view (one-half of the unit cell along z-axis) of calcite with the same iso-surface level is shown in figure 4.5.3 in order to visualize the bonding nature between the atoms. Figure 4.5.4 shows all the trigonal planar CO_3 groups consisting of carbon atom surrounded by three oxygen atoms. Another view (one-half of the unit cell along z-axis) with the trigonal planar CO_3 groups of atom in calcite structure is shown in figure 4.5.5 The trigonal planar CO_3 groups are constrained to be parallel and constitute a plane perpendicular to c-axis in the unit cell of calcite and the trigonal planar CO_3 groups thus formed within the unit cell of calcite are more symmetrical. This is a typical feature of carbonate crystal structure, determining strong anisotropy of their physical properties. Each Ca atom is coordinated by six O atoms at the vertices of an octahedron. The bond length between C–O in the planar CO_3 and Ca–O in octahedral structure of calcite is found to be 1.2647 Å and 2.3695 Å respectively. The 3D view (one-fourth of the unit cell along z-axis) with the trigonal planar CO_3 group of atoms in the structure of calcite considering the iso-surface level of 1.3 e/Å^3 is shown in figure 4.5.6. The bonding nature between C–O atoms can be obviously visualized in the trigonal planar CO_3 group. Ca atom is in octahedral coordination with -3 symmetry is shown in figure 4.5.7. Each O atom is shared between two octahedra and also forms one corner of an equilateral triangle, perpendicular to the c-axis, with the carbon atom at its centre in the unit cell of calcite as shown (one-half of the unit cell of calcite along z-axis) in figure 4.5.8. These facts cause the groups to align themselves between the layers of cations in such a way that each O in the group is attached to a cation from above and below with equal bonding distance that balance each other.

3D outlook of the unit cell of calcite with the (001) plane is shown in figure 4.5.9 and figure 4.5.10 shows the view of the same (001) plane which is passing through the Ca atoms in the perpendicular direction. Figure 4.5.11 shows the 2D electron density distribution corresponding to the (001) plane. The contour range is from 0.0 e/Å^3 to 3.0 e/Å^3 and the interval is 0.12 e/Å^3. The 3D view of the unit cell of calcite with the (001) plane at a distance of 1.505 Å away from the origin of the cell is shown in figure 4.5.12.

The two-dimensional electron density distribution on this plane is shown in figure 4.5.13. The contour range is from 0.0 e/Å^3 to 3.0 e/Å^3 and the interval is 0.17 e/Å^3. The covalent bonding nature between oxygen and carbon within the trigonal planar CO_3 group can be visualized through the contour lines in both the figures 4.5.12 and 4.5.13.

Figure 4.5.14 shows the 3D view of the unit cell of calcite with the (010) plane and the 2D electron density distribution on that plane with the contour range from 0.0 e/Å^3 1.8 e/Å^3 with the interval 0.18 e/Å^3 is shown in figure 4.5.15. The 3D view of the unit cell of calcite with the (101) is shown in figure 4.5.16. The 2D electron density distribution on that plane with the contour range from 0.0 e/Å^3 to 2.0 e/Å^3 and contour interval 0.3 e/Å^3 is shown the figure 4.5.17. Figure 4.5.18 shows the unit cell of calcite with the (110) plane. The 2D electron density distribution on that plane with the contour range from 0.0 to 1.8 e/Å^3 and contour interval 0.24 e/Å^3 is shown in figure 4.5.19.

Table 4.5.1 Parameters from MEM refinement of CaCO₃

Parameter	Value
Number of cycles	3051
Lagrange parameter (λ)	0.0040
Number of electrons/unit cell (F_{000})	300
R_{MEM} (%)	1.1443
wR_{MEM} (%)	1.8522

R_{MEM} - Reliability index from MEM refinement

$_wR_{MEM}$ - Weighted reliability index from MEM refinement

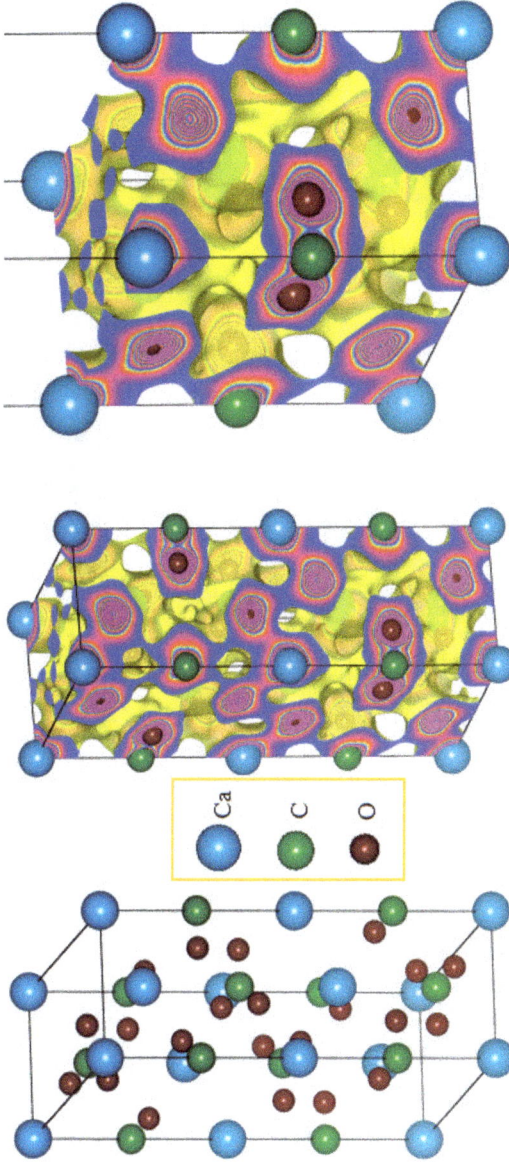

Figure 4.5.1 Unit cell of CaCO₃.

Figure 4.5.2 3D view of electron density in the unit cell of CaCO₃ (iso-surface level: 0.15 e/Å³).

Figure 4.5.3 3D view (half the unit cell along z-axis) of electron density in the unit cell of CaCO₃ (iso-surface level: 0.15 e/Å³).

Figure 4.5.6 3D representation (one-eighth of the unit cell along z-axis) of trigonal planar CO₃ group in the unit cell of CaCO₃ (iso-surface level: 1.2 e/Å³).

Figure 4.5.5 3D representation (one-half of the unit cell along z-axis) of trigonal planar CO₃ groups in the unit cell of CaCO₃.

Ca
C
O

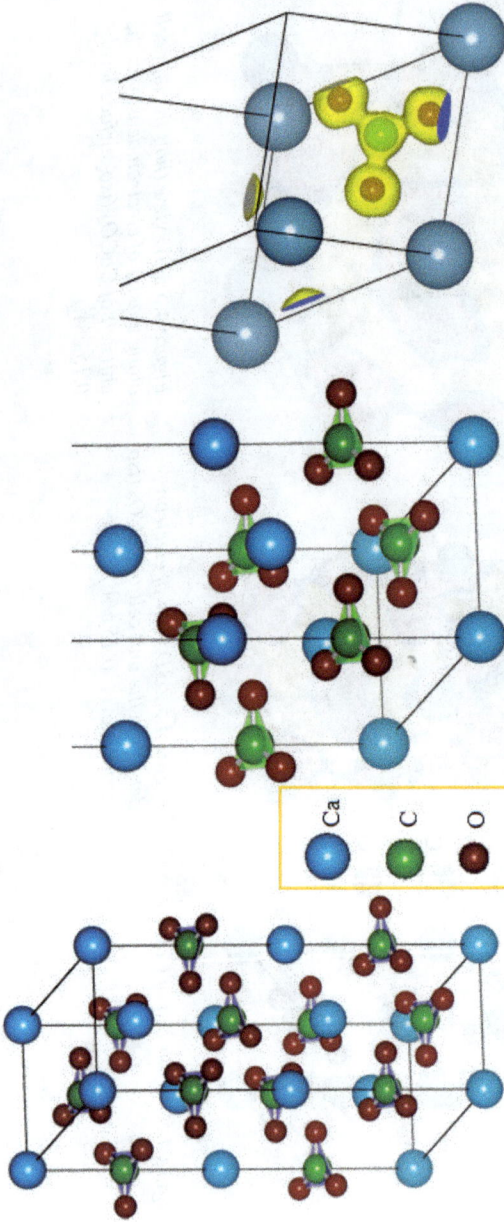

Figure 4.5.4 Unit cell of CaCO₃ containing trigonal planar CO₃.

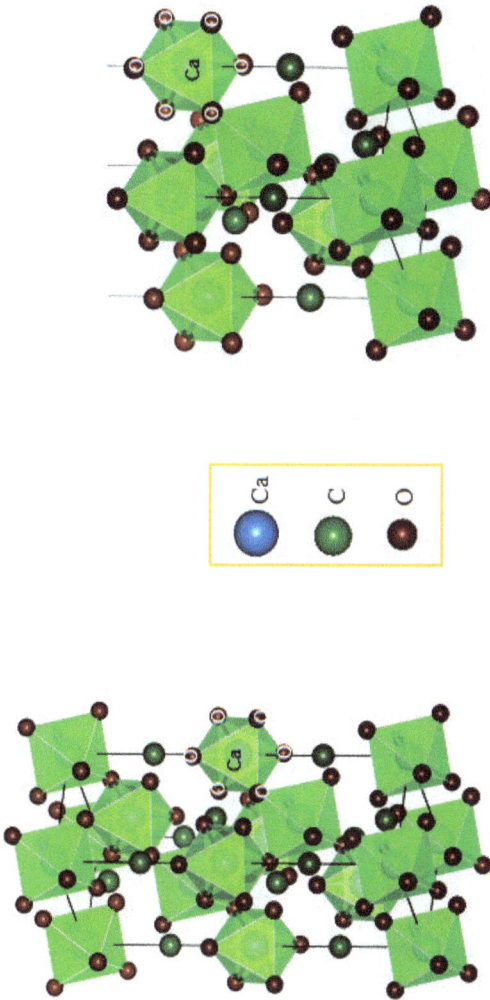

Figure 4.5.7 3D view of octahedral structure of Ca with O
atoms in the unit cell of CaCO₃.

Figure 4.5.8 Unit cell of calcite (one-eighth of the unit cell
along z-axis) containing CaO₆ octahedra and CaO₄
tetrahedra in the unit cell of CaCO₃.

Figure 4.5.11 2D MEM electron density distribution on (001) plane passing through Ca atoms in the unit cell of $CaCO_3$ (Contour interval: 0 to 3 step size of 0.12 $e/Å^3$).

Figure 4.5.10 3D electron density of calcite with (001) plane passing through Ca atoms in the perpendicular direction in the unit cell of $CaCO_3$.

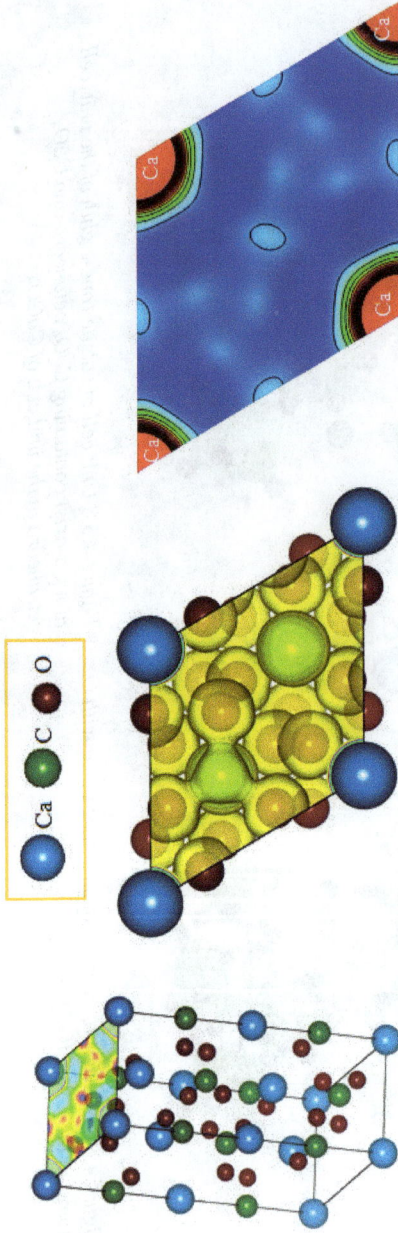

Figure 4.5.9 The (001) plane passing through Ca atoms in $CaCO_3$.

Figure 4.5.13 2D MEM electron density distribution on (001) plane at a distance of 1.505 Å away from the origin (Contour interval: 0 to 3 step size of 0.17 e/Å³).

Ca	●
C	●
O	●

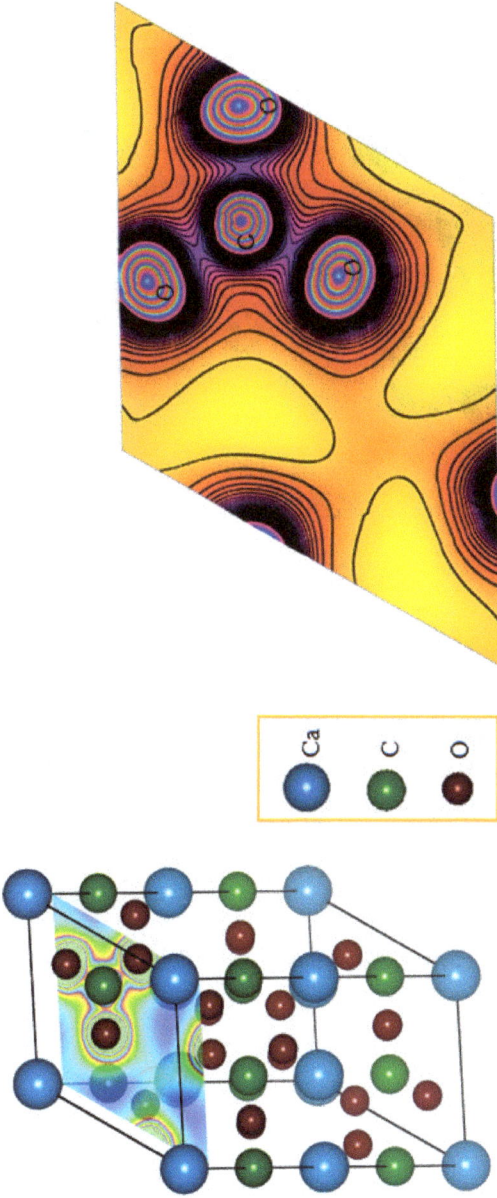

Figure 4.5.12 3D view with the (001) plane at a distance of 1.505 Å away from the origin in the unit cell of CaCO₃.

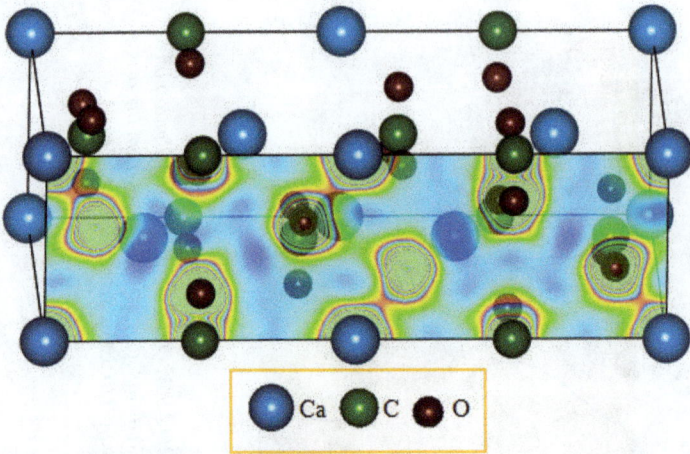

Figure 4.5.14 The (010) plane passing through Ca, C and O atoms in the unit cell of $CaCO_3$.

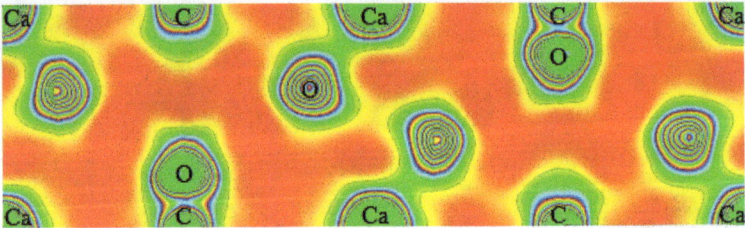

Figure 4.5.15 2D MEM electron density distribution on (010) plane of $CaCO_3$ (Contour interval: 0 to 2 step size of $0.3 e/Å^3$).

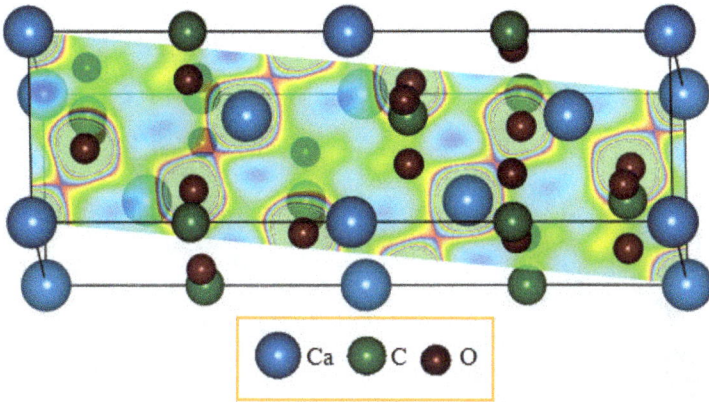

Figure 4.5.16 The (101) plane passing through Ca and O atoms in the unit cell of CaCO₃.

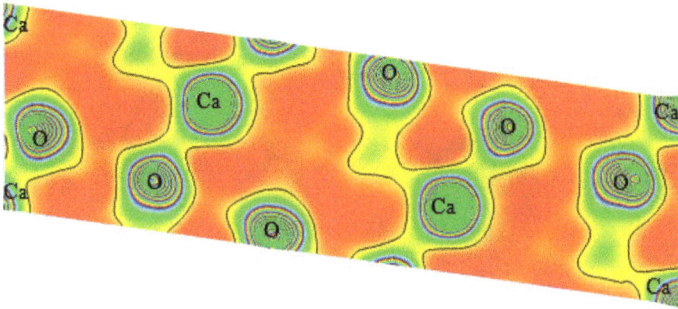

Figure 4.5.17 2D MEM electron density distribution on (101) plane of CaCO₃ (Contour interval: 0 to 3 step size of 0.12 e/$Å^3$).

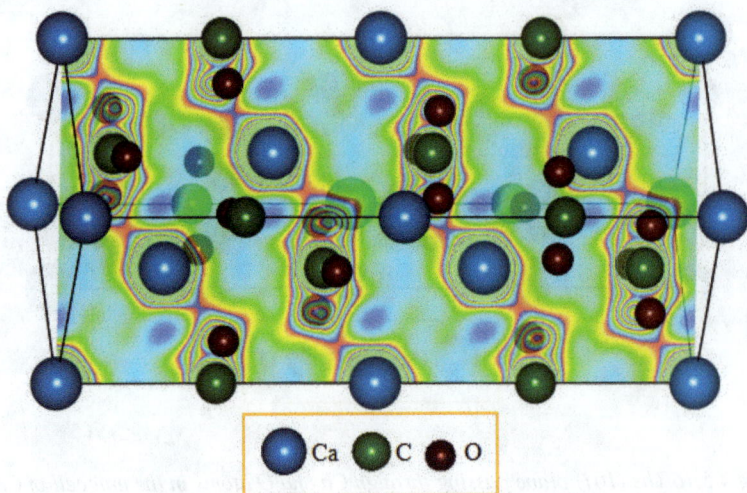

Figure 4.5.18 3D view of (110) plane passing through Ca, C and O atoms in the unit cell of $CaCO_3$.

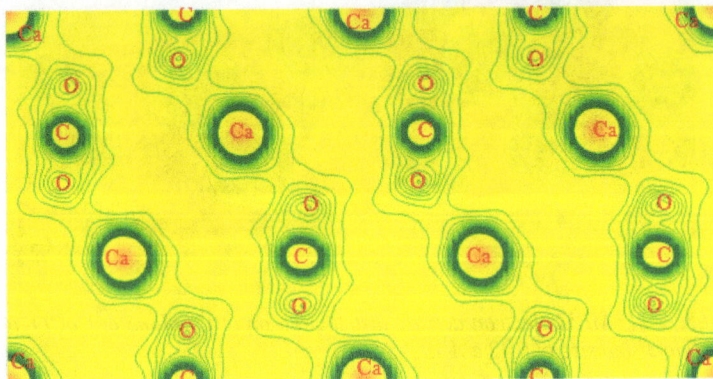

Figure 4.5.19 2D MEM electron density distribution on (110) plane of $CaCO_3$ (Contour interval: 0 to 1.8 step size of 0.24 $e/Å^3$).

The qualitative understanding and the analysis can be done by visualizing the one dimensional profiles of the charge density along the bonding directions. Hence, the one-dimensional profiles of electron density are drawn along Ca–O and C–O directions and are shown in figures 4.5.20 and 4.5.21 respectively. The numerical values of the mid bond electron densities between different atoms of calcite from one-dimensional MEM analysis are given in table 4.5.2. The ionic and covalent nature of bonding is revealed in Ca–O and C–O respectively. The first atom has been considered to be in the origin in the above mentioned one-dimensional electron density profiles.

Table 4.5.2 MEM mid-bond electron densities between different atoms of calcite obtained from 1D electron density profiles.

Bonding direction	Position (Å)	Mid-bond electron density (e/Å^3)
Ca–O	2.3695	0.4994
C–O	1.2647	1.7856

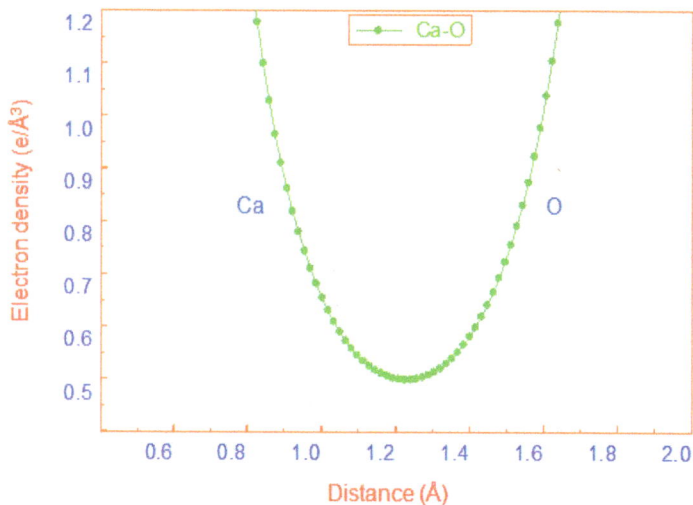

Figure 4.5.20 1D variation of electron density along Ca–O bond.

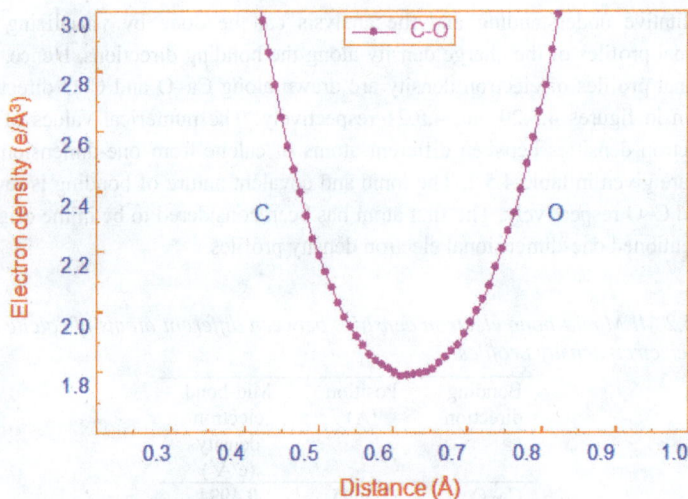

Figure 4.5.21 1D variation of electron density along C–O bond.

4.6 Electron density analysis of $Ca_{1-x}Yb_xF_2$

It is clear from the earlier studies of several researchers that the probability based maximum entropy method (MEM) (Collins, 1982) theory is a powerful and elegant tool for studying the electron density distributions including low-density regions. The refined MEM parameters of $Ca_{1-x}Yb_xF_2$ (x = 0, 0.03, 0.06, 0.09 and 0.12) have been given in table 4.6.1.

Figure 4.6.1(a) represents the individual atoms imposed on the unit cell of calcium fluoride. The three-dimensional charge density distribution constructed using the MEM (Collins, 1982) model is presented with the same iso-surface level of 0.3 e/$Å^3$ in the unit cell for the prepared samples of $Ca_{1-x}Yb_xF_2$ (x = 0.00, 0.03, 0.06, 0.09 and 0.12) along with a ball and stick model as shown in figures Figure 4.6.1(b-f) respectively.

The bonding between Ca and F atoms are clearly visualized through the electron density distribution in the unit cell of CaF_2. Figure 4.6.1(b) shows the visualization of electron clouds in the unit cell of CaF_2 without doping of Yb. The increased concentration of Yb at 3% gets more charges at the mid-point between Ca and F atom as shown in figure 4.6.1(c). At this Yb doping concentration major charge ordering takes place and the system becomes more ionic. The formation of defect enables the crystal with the fluorite structure to accommodate a large number of anion interstitials. At low rare earth concentration, Yb^{3+} is substituted in the host lattice of divalent cation and charge compensation is effected by the anion interstitials and it leads to the cluster formation as the concentration increases (Corish et al., 1982). Petit et al., (2008) have reported that there are three doping regions for $Yb:CaF_2$ single crystals. Lyberis et al., (2001) have reported that the optimal composition to achieve the best transparent ceramic is close to 5 at% in ytterbium doped CaF_2. The charge density clouds are isolated when doping of Yb at 6% as shown in figure 4.6.1(d). Figure 4.6.1(d) reveals a unique situation where the charges spread themselves from their respective atomic positions to be in the valence region. The mid bond electron density between the atoms increases further with the increased doping concentration of Yb at 9% and 12% as shown in figures 4.6.1(e) and 4.6.1(f) respectively.

Figure 4.6.2(a) shows the (110) plane passing through the unit cell of CaF_2. Figures 4.6.2(b-f) show the 2D charge density distribution corresponding to the (110) plane of $Ca_{1-x}Yb_xF_2$ (x = 0.00, 0.03, 0.06, 0.09 and 0.12) respectively. The contour range is from 0.0 to 4.0 e/Å^3 and the contour interval is 0.22 e/Å^3. Figure 4.6.2(b) shows the two dimensional electron density distribution map for undoped CaF_2. In figure 4.6.2(c), the contribution of Yb^{3+} at 0.03 to the Ca matrix seems to increase the charge density when compared with undoped CaF_2. The 2D charge density distribution shown in figure 4.6.2(d) for Yb concentration of 6% indicates the depletion of charge density in the valence region which results in the decrease of bond critical point (BCP) value. The increase in Yb at 0.09 and 0.12 increases the strength of bonding as shown in figures 4.6.2(e) and 4.6.2(f) respectively. The magnitude of the valence charges can also be quantified by understanding the 1D charge density profiles drawn along the bonding directions. It clearly shows the Ca-F bond with ionic character which is confirmed by the bond critical point (BCP) values from one dimensional profile between Ca and F atom as shown in figure 4.6.3.

Table 4.6.1 Parameters from MEM refinements of $Ca_{1-x}Yb_xF_2$ materials.

Parameters	Composition of Yb				
	0%	3%	6%	9%	12%
Number of cycles	836	1684	1030	1029	1070
Lagrange parameter (λ)	0.002	0.002	0.002	0.002	0.002
R_{MEM} (%)	1.44	1.11	1.25	1.21	1.14
wR_{MEM} (%)	1.37	1.14	1.28	1.23	1.18

R_{MEM} - Reliability index from MEM refinement

$_wR_{MEM}$ - Weighted reliability index from MEM refinement

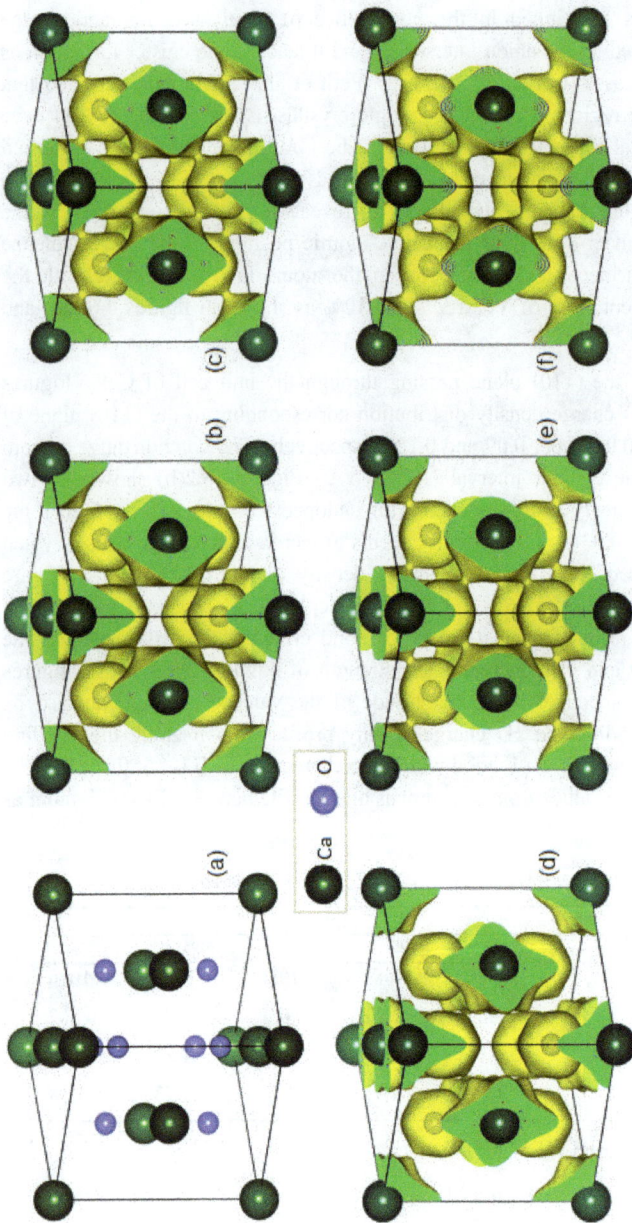

Figure 4.6.1(a) 3D view of the unit cell of CaF$_2$; Ca$_{1-x}$Yb$_x$F$_2$ [(b) x = 0.00 (c) x = 0.03 (d) x = 0.06 (e) x = 0.09 and (f) x = 0.12)] (iso-surface: 0.3 e/Å^3).

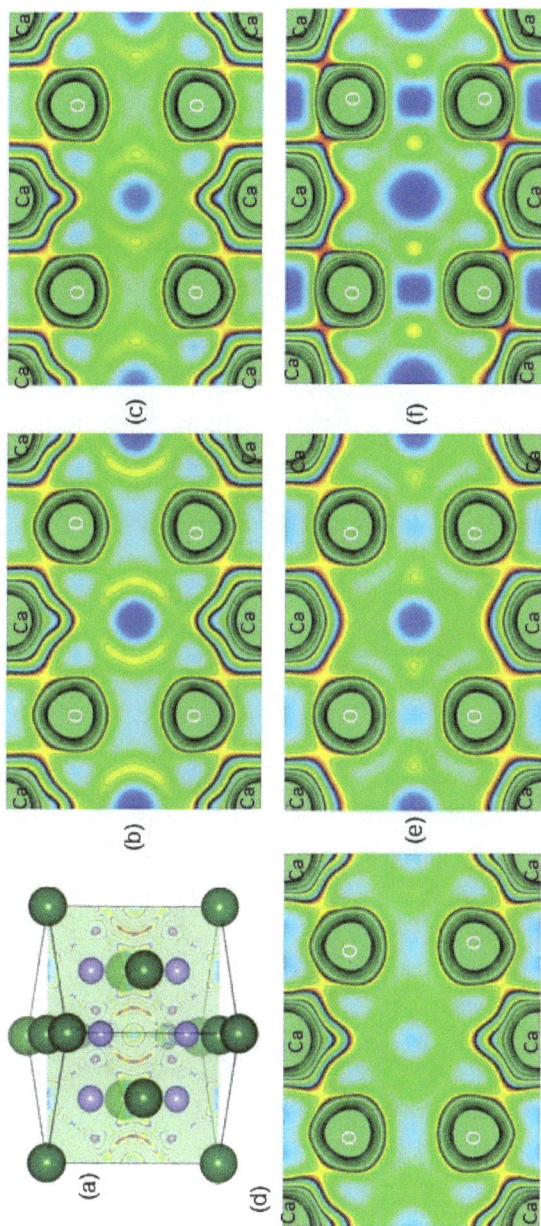

Figure 4.6.2(a) View of (001) plane passing through the unit cell of CaF₂; 2D electron density map drawn on (001) plane of Ca₁₋ₓYbₓF₂ [(b) x = 0.00 (c) x = 0.03 (d) x = 0.06 (e) x = 0.09 and (f) x = 0.12)] (Contour interval: 0 to 4 step size of 0.22 e/ų).

The one-dimensional electron density profiles of $Ca_{1-x}Yb_xF_2$ (x = 0.00, 0.03, 0.06, 0.09 and 0.12) are shown in figure 4.6.3. The electron densities along the bond critical points (BCP) between Ca and F for $Ca_{1-x}Yb_xF_2$ (x = 0.00, 0.03, 0.06, 0.09 and 0.12) from one-dimensional MEM analysis are given in table 4.6.2.

Table 4.6.2 MEM mid-bond electron densities between Ca–F bonds of $Ca_{1-x}Yb_xF_2$ materials obtained from 1D electron density profiles.

Composition x	Position (Å)	Electron density $(e/Å^3)$
0.00	1.4393	0.3046
0.03	1.4393	0.3220
0.06	1.4055	0.2922
0.09	1.4055	0.3157
0.12	1.3930	0.4013

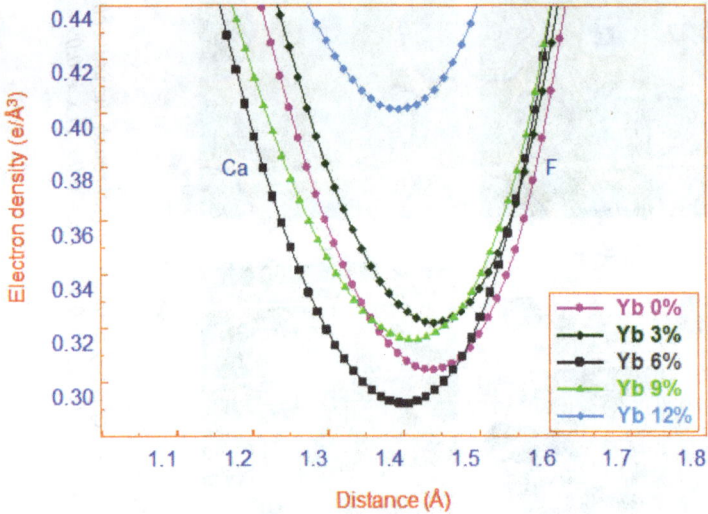

Figure 4.6.3 1D electron density profiles of Ca–F atoms in the unit cell of $Ca_{1-x}Yb_xF_2$ (x = 0.00, 0.03, 0.06, 0.09 and 0.12).

4.7 Electron density analysis of Al_2O_3, $Cr:Al_2O_3$ and $V:Al_2O_3$

The qualitative and quantitative electron distribution of the atoms in crystalline materials of Al_2O_3, $Cr:Al_2O_3$ and $V:Al_2O_3$ were studied using maximum entropy method (MEM) (Collins, 1982). The MEM parameters of Al_2O_3, $Cr:Al_2O_3$ and $V:Al_2O_3$ have been given in table 4.7.1.

Figure 4.7.1 represents the three-dimensional electron density distribution in the unit cell of Al_2O_3. Figure 4.7.2 represents the (100) plane passing through the unit cell of Al_2O_3. The two-dimensional electron density distribution on the (100) plane for Al_2O_3, $Cr:Al_2O_3$ and $V:Al_2O_3$ are represented in figures 4.7.3, 4.7.4 and 4.7.5 respectively. The contour range is from 0.0 to 1.0 e/$Å^3$ and the contour interval is 0.09 e/$Å^3$. The values of mid-bond electron densities from the one-dimensional electron density profiles of Al_2O_3, $Cr:Al_2O_3$ and $V:Al_2O_3$ are given in table 4.7.2. The one-dimensional electron density profiles of Al_2O_3, $Cr:Al_2O_3$ and $V:Al_2O_3$ are shown in figures 4.7.6, 4.7.7 and 4.7.8 respectively.

Table 4.7.1 Parameters from MEM refinements of Al_2O_3, $Cr:Al_2O_3$ and $V:Al_2O_3$.

Parameter	Al_2O_3	$Cr:Al_2O_3$	$V:Al_2O_3$
Number of cycles	75	1068	290
Lagrange parameter (λ)	0.087	0.0694	0.1944
Number of electrons/unit cell(F_{000})	300	313	312
R_{MEM} (%)	0.0225	0.0452	0.0355
wR_{MEM} (%)	0.0298	0.0339	0.0281

R_{MEM} - Reliability index from MEM refinement

$_wR_{MEM}$ - Weighted reliability index from MEM refinement

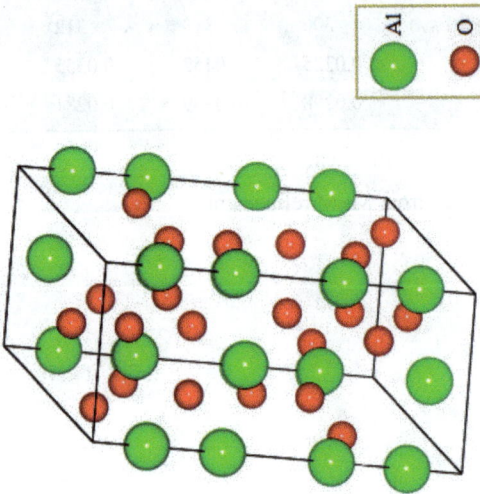

Figure 4.7.2 3D view in the unit cell of Al₂O₃ with the (100) plane.

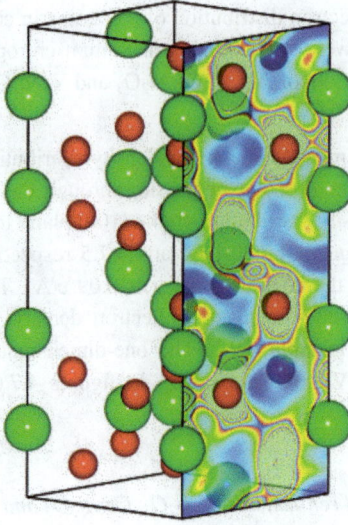

Al

O

Figure 4.7.1 3D view in the unit cell of Al₂O₃.

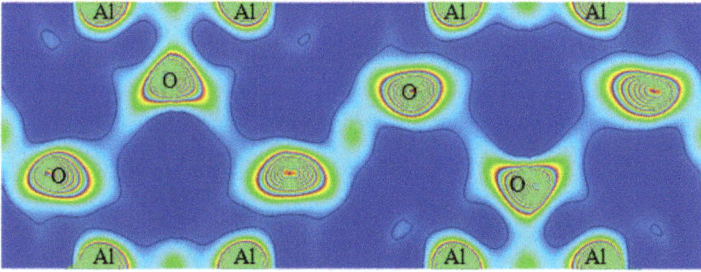

Figure 4.7.3 2D electron density map drawn on (100) plane of Al_2O_3 (Contour interval: 0 to 1 step size of 0.1 e/$Å^3$).

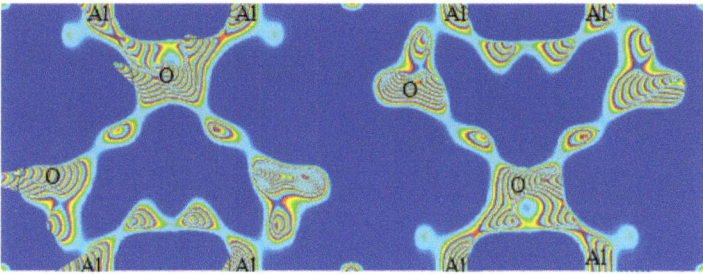

Figure 4.7.4 2D electron density map drawn on (100) plane of $Cr:Al_2O_3$ (Contour interval: 0 to 1 step size of 0.1 e/$Å^3$).

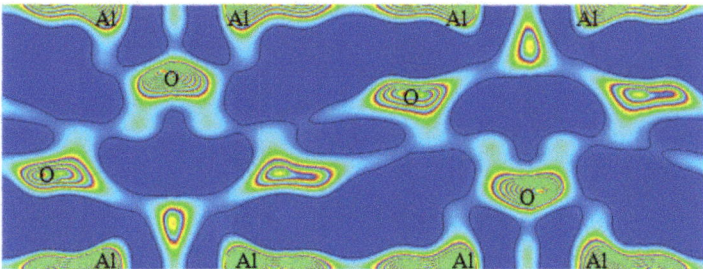

Figure 4.7.5 2D electron density map drawn on (100) plane of $V:Al_2O_3$ (Contour interval: 0 to 1 step size of 0.1 e/$Å^3$).

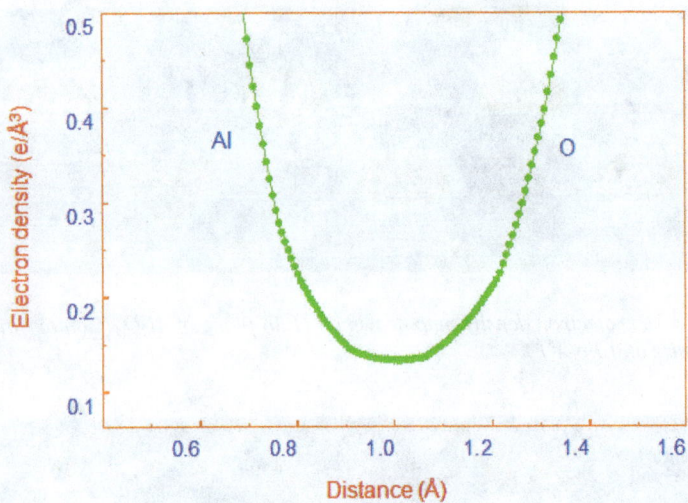

Figure 4.7.6 1D electron density profile of Al_2O_3.

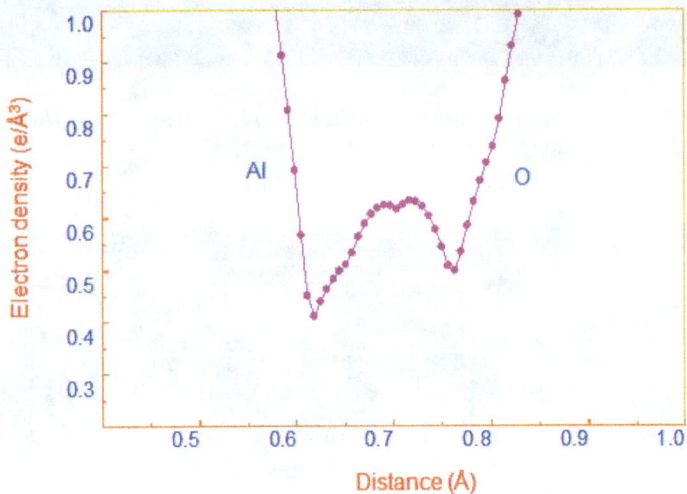

Figure 4.7.7 1D electron density profile of $Cr:Al_2O_3$.

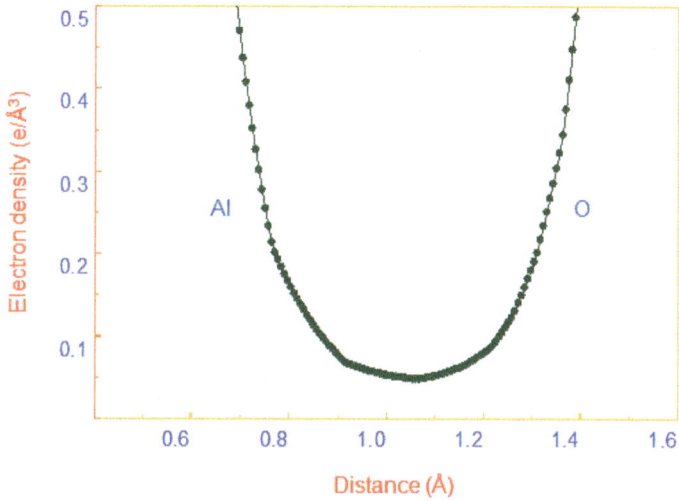

Figure 4.7.8 1D electron density profile of V:Al$_2$O$_3$.

Table 4.7.2 Mid-bond electron densities of Al$_2$O$_3$, Cr:Al$_2$O$_3$ and V:Al$_2$O$_3$ from 1D electron density profile.

Material	Position (Å)	Electron density (e/Å^3)
Al$_2$O$_3$	1.0067	0.1326
Cr:Al$_2$O$_3$	0.6182	0.4119
	0.7037	0.6165
	0.7629	0.5005
V: Al$_2$O$_3$	1.0677	0.0485

4.8   Conclusion

The electron density distribution analysis have been done for all the chosen non-linear optical materials, PbMoO$_4$, LiNbO$_3$, Ce:Gd$_3$Ga$_5$O$_{12}$, CaCO$_3$, Yb:CaF$_2$ and Al$_2$O$_3$,Cr:Al$_2$O$_3$ & V:Al$_2$O$_3$, using statistical approach, maximum entropy method.

The electron density distribution of $PbMoO_4$ has been mapped using the MEM electron density values obtained through refinements. The nature of bonding and the interaction due to different atoms has been analysed through 3D iso-surfaces, two dimensional electron density distribution contour maps and the quantitative electron densities from the one dimensional profiles. The hybridization of 4d state of Mo atom and 6s state of Pb atom with the 2p state of the oxygen atom leads to the essential properties of $PbMoO_4$ material.

The electron density distribution of $LiNbO_3$ has been elucidated as 3D iso-surfaces, 2D electron density maps on various planes including those with prominent bonding features and 1D electron density profiles along various atoms of interest in the unit cell. The lowest electron density arises between Li–O interactions than Li–Li and O–O interactions. The mid bond electron densities from the one-dimensional electron density profiles confirm that the bonding between Li–O bonding is of more ionic character and the Nb–O bond is mixed covalent with ionic character. The Li–O interaction plays a key role in the physical properties of $LiNbO_3$. Li–Li and O–O interactions are being used to stabilize the system.

The variation of bonding feature has been discussed through maximized entropy method in Ce doped gadolinium gallium garnet ($Ce:Gd_3Ga_5O_{12}$). The electron density distribution has been analysed through three, two dimensional electron density maps and one dimensional profiles in the unit cell. For Gd–O bond in dodecahedron, the mid bond electron density ranges from 0.2962 e/$Å^3$ to 0.5043 e/$Å^3$, which reveals the ionic with partly covalent nature and for Ga(1)–O bond in octahedra, the mid bond electron density ranges from 0.6504 e/$Å^3$ to 0.8868 e/$Å^3$, revealing the covalent nature. Large influence of the charge density distributions plays an important role in optical property which may be the prominent key factor for the luminescence properties.

The electron density distribution analysis in the unit cell of $CaCO_3$ has been done using maximum entropy method. The electron density maps give qualitative as well as quantitative picture of the electronic structure in the unit cell of $CaCO_3$. The two dimensional electron density distributions show the covalent along with ionic nature existing between Ca–O atoms and perfect covalent nature existing between C–O atoms. The mid bond electron densities from one-dimensional electron density profiles confirm the existence of covalent along with ionic and perfect covalent character between Ca–O and C–O atoms respectively.

The electron density distributions of $Ca_{1-x}Yb_xF_2$ (x = 0.00, 0.03, 0.06, 0.09 and 0.12) material has been elucidated as 3D iso-surfaces, 2D electron density maps and 1D electron density profiles in the unit cell of CaF_2 using maximum entropy method (MEM).

The electronic charge around Ca lattice site gets modified due to the incorporation of trivalent Yb atom. At Yb 6%, the mid bond electron density decreases and it behaves unusual one.

The electron density distribution of pure and doped Al_2O_3 has been analyzed using maximum entropy method. The results indicate that it is possible to distinguish the doping effects in Al_2O_3 through the charge density analysis. The redistribution of charge density due to the inclusion of Cr and V on host matrix of Al_2O_3 is clearly visualized through 3D electron density iso-surfaces, the 2D electron density contour maps on particular crystallographic plane and the 1D electron density profiles. The mid bond electron density for Al-O bond in pure Al_2O_3 is 0.1326 $e/Å^3$ which reveals the mixed ionic and covalent character. The mid bond electron density for Al–O bond in $Cr:Al_2O_3$ is 0.6165 $e/Å^3$ which reveals more covalent with partly ionic character. The mid bond electron density for Al–O bond in $V:Al_2O_3$ is 0.0485 $e/Å^3$ which reveals predominant ionic nature.

References

[1]   Bricogne G and Gilmore C. J, Acta Cryst., Vol. A46, pp. 284, 1990. https://doi.org/10.1107/S0108767389012882

[2]   Collins D, Nature, Vol. 298, pp. 49, 1982. https://doi.org/10.1038/298049a0

[3]   Corish J., Catlow C., Jacobs P. W. M and Ong S. H., Phys Rev B, Vol. 25, pp. 6425, 1982. https://doi.org/10.1103/PhysRevB.25.6425

[4]   Gilmore C. J., Acta Crystallogr. A, Vol. 52, pp. 561, 1996. https://doi.org/10.1107/S0108767396001560

[5]   Izumi F and Dilanian R. A, Recent Research Developments in Physics, Vol. 3, Part II, Transworld Research Network, Trivandrum, 2002

[6]   Lyberis A., Stevenson A. J., Suganuma A., Ricaud S., Druon F., Herbst F., Vivien D., Gredin P and Mortier M., Optical Materials, Vol. 34, pp. 965, 2001. https://doi.org/10.1016/j.optmat.2011.05.036

[7]   Momma K and Izumi F, Journal of Applied Crystallography, Vol. 44, pp. 1272, 2011. https://doi.org/10.1107/S0021889811038970

[8]   Petit V., Doualan J. L, Camy P., Portier X and Moncorge R., Phys. Rev. B, Vol. 78, pp. 1, 2008

[9]   Rietveld H. M, Journal of Applied Crystallography, Vol. 2, pp. 65, 1969. https://doi.org/10.1107/S0021889869006558

[10]   Saka T and Kato N., Acta Crystallogr. A, Vol. 42, pp. 469, 1986.
https://doi.org/10.1107/S0108767386098860

[11]   Sakata M and Sato M, Acta Crystallogr. A Vol. 42, pp. 263, 1990.
https://doi.org/10.1107/S0108767389012377

[12]   Saravanan R, Ono Y, Ohno M, Isshiki K, Ohno K and Kajitani T, J. Phys. Chem.
Solids, Vol. 64, pp.51, 2003. https://doi.org/10.1016/S0022-3697(02)00209-3

[13]   Yamamoto K, Takahashi Y, Ohshima K, Okamura F and Yukino B. K, Acta
Crystallogr. A, Vol. 52, pp. 606, 1996.
https://doi.org/10.1107/S0108767396001845

[14]   Yamamura S., Takata M., Sakata M and Sugawara Y., J. Phys. Soc. Japan, Vol.
67, pp. 4124, 1998. https://doi.org/10.1143/JPSJ.67.4124

[15]   Sczancoski J. C., Bomio M. D. R., Cavalcante L. S., Joya M. R., Pizani P. S.,
Varela J.A., Longo S., Siu Li M and Andrés J. A., J. Phys. Chem. C, Vol. 14, pp.
113, 2009

Chapter 5

Analysis of Pair Distribution Function of Non-Linear Optical Materials

Abstract

Chapter V deals with the study of the inter-atomic ordering i.e., the local structure ordering of the selected non-linear optical materials. The pair distribution function (atomic correlation function), is computed for selected non-linear optical materials of the present work.

Keywords

Local Structure, Pair Correlation Function, Inter Atomic Order, Crystalline, Amorphous

Contents

5.1   Introduction

The study of the local atomic arrangement of materials is one of the important features for the understanding and possible prediction of interesting macroscopic properties of modern materials. Nowadays, crystallographic analysis methods are able to provide refinement of crystal structures. In general, conventional determination of crystal structure is based on the analysis of the intensities and positions of Bragg reflections which allows only the determination of the long-range average structure of the crystal. It is possible to obtain the information about bond length distribution both static, thermal and correlated atomic thermal motion (Petkov et al., 1999) from the pair distribution function (PDF) peak width. Rietveld (1969) structural refinement method is usually performed for the determination of crystal structure of powder diffraction data. In reality, all the compounds are not single crystals or well crystallized powders. Amorphous compounds have shown the limits of the crystallographic approach since it only yields the average structure of the material as it is based on the analysis of Bragg intensities. On the other hand, deviations from the average structure result in diffuse scattering which in general contains two-body atomic correlations (Welberry and Butler, 1995; Frey, 1995) and thus holds the key to local structure of the material. X-ray absorption fine structure (XAFS) is the first experimental options to study inter atomic arrangement of materials (Sayers et al., 1971; Filipponi et al., 1995; Stumm, 1989; Rehr and Albers, 2000; deGroot, 2001; Lytle, 1999). Pair distribution function (PDF) is one of the alternate methods to reveal local structure from the powder X-ray diffraction or the neutron diffraction data. The pair distribution function (PDF) is obtained by Fourier transform of total scattering of powder diffraction pattern. For a long time, pair distribution function (PDF) method is well known in the field of studies of short-range order in liquids, glasses and amorphous solids. In recent years, pair distribution function (PDF) method is also applied equally well to crystalline materials too if they are in the form of fine powders or polycrystalline aggregates (Toby and Egami, 1992; Billinge and Egami, 1993; Billinge and Kanatzidis, 2004). Quantitative structural information on nanometre length scales can be obtained by fitting a model directly to the PDFfit (Proffen and Billinge, 1999) based on the equation,

$$G(r) = 4\pi r[\rho(r) - \rho_0] = \frac{2}{\pi} \int \vec{Q}[S(\vec{Q}) - 1]sin(\vec{Q}r) d\vec{Q} \tag{5.1}$$

where $G(r)$ is the atomic pair distribution function and $\rho(r)$ corresponds to the (atomic) number density at a distance r from the average atom. PDFgetX (Jeong et al., 2001) is a program used to obtain the observed atomic pair distribution function (PDF) from X-ray powder diffraction data. The graphical software PDFgui (Farrow et al., 2007) is used, which is a graphical environment for PDF fitting to plot the observed and calculated PDF.

In the present work, the pair distribution function (PDF) approach has been utilized for the study of the selected non-linear optical materials, lead molybdate ($PbMoO_4$) and lithium niobate ($LiNbO_3$). The PDF for $PbMoO_4$ and $LiNbO_3$ have been refined using PDFfit (Proffen and Billinge, 1999). The bond lengths deduced from the PDF analysis for the samples $PbMoO_4$ and $LiNbO_3$ have been compared with those calculated using software program GRETEP [Grenoble Thermal Ellipsoids Plot Program is a Windows interactive program] (Jean Laugier and Bernard Bochu, 2002).

5.2   Pair distribution function analysis of PbMoO$_4$

In order to understand the local structure, the powder X-ray data of lead molybdate (PbMoO$_4$) has been utilized for the analysis of pair distribution function (PDF). The PDF for PbMoO$_4$ has been refined using PDFgui (Farrow et al., 2007) and the comparison of observed and calculated PDF's has been made and analyzed.

The refined parameters of lead molybdate (PbMoO$_4$) using pair distribution function (PDF) are given in table 5.2.1. The nearest neighbour atom distances obtained from the PDF analysis are given in table 5.2.2. In our work, the upper limit for the Fourier transform of the data was set until $Q_{max} = 6.5$ Å^{-1}. The fitted observed and calculated pair distribution functions of PbMoO$_4$ are shown in figure 5.2.1. In this figure, r(Å) represents the atomic distance from the origin and G(r) represents the reduced pair distribution function. Figure 5.2.1 shows the PDF peaks representing to the interaction between Pb–Pb, Pb–O, Mo–Mo, Mo–O and O–O atomic pairs. PDF peaks up to a distance of 20 Å are suitably indexed as shown in figure 5.2.1. It is worthy to note that the agreement between the observed and calculated PDF's is very good with small error function. Furthermore, the high Q synchrotron data is essential for this type of analysis on short range atomic correlation study and local structure of materials (Jeong et.al., 2001; Neder and Proffen, 2009; Egami, 1990). In spite of this factor, an attempt has been successfully made to use the X-ray powder data for this analysis.

Table 5.2.1 The refined parameters from pair distribution function of PbMoO$_4$.

Parameter	PbMoO$_4$
Data range (Å)	0.0 – 25.0
Refinement range (Å)	2.0 – 20.0
Change in R	0.0019

R - Reliability index

Table 5.2.2 Nearest neighbour distances from PDF analysis of PbMoO$_4$.

Atom pair	Inter-atomic distances from PDF analysis (Å)		*Calculated inter-atomic distances (Å)
	Observed	Calculated	
Pb–Pb	3.98	3.98	3.593
Pb–O	6.80	6.86	6.983
Mo–Mo	8.48	8.52	8.499
O–O	10.50	10.54	10.576
Pb–O	13.46	13.50	13.417
Mo–O	15.02	15.06	15.035
Mo–O	16.50	16.40	16.485
Pb–Pb	17.10	17.14	17.120
Mo–Mo	18.34	18.38	18.373
Mo–Pb	20.18	20.16	19.764

*GRETEP for Windows (Jean Laugier and Bernard Bochu, 2002)

From table 5.2.2, it can be observed that the difference in the observed and calculated nearest neighbour distances turns out to be very small. These differences are not too high considering the fact that we have utilized only data with limited Q values. There may be local undulations in the structure of this system that will lead to slightly higher neighbour differences which correspond to lattice repeat distances. Also, the local disorder is expected to affect the repeat distances, because of the diffuse scattering which does not have sharp, single pointed X-ray diffraction phenomenon. Hence, it is reflected in the local structural analysis.

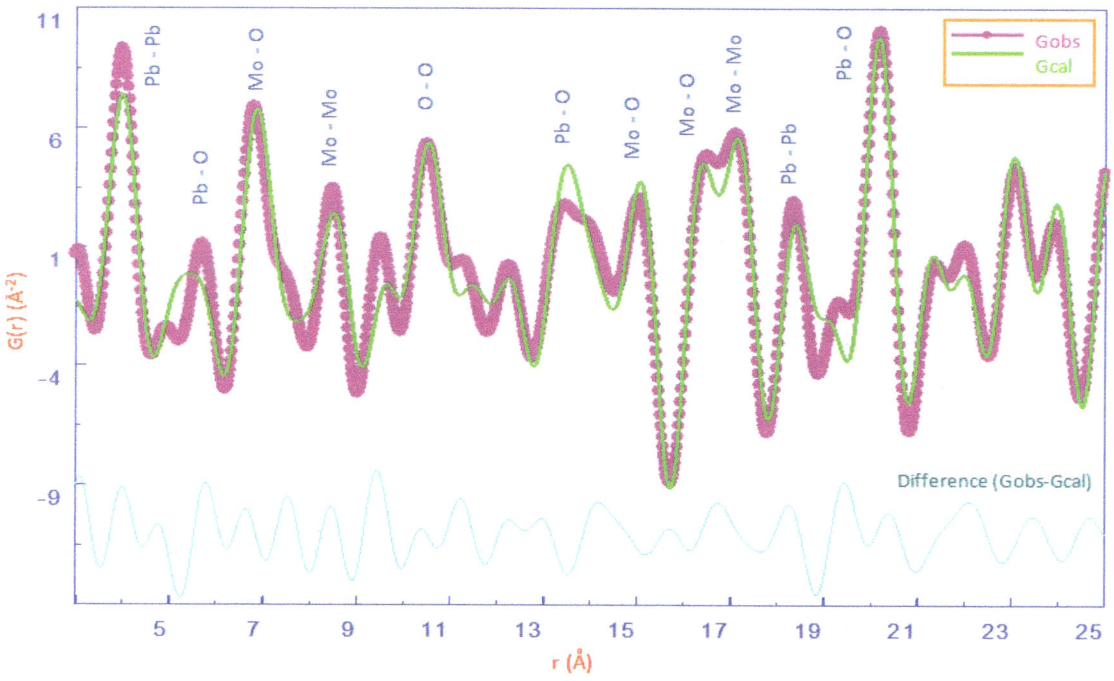

Figure 5.2.1 The fitted observed and calculated pair distribution function of PbMoO₄.

5.3 Pair distribution function analysis of LiNbO₃

The powder X-ray data of LiNbO$_3$ has been employed for the analysis of pair distribution function. The observed PDF's have been obtained from the experimental powder XRD data, using the software package PDFgetX (Jeong et al., 2001). PDFgetX can help reduce the raw data into a more convenient format from which the analysis can be done within the desired range (r). In our work, the upper limit for the Fourier transform of the data was set until Q_{max} = 6.5 Å^{-1}. After the preliminary data reduction, we have used PDFgui (Farrow et al., 2007) for fitting the theoretical structure models with experimental pair distribution functions in a refinement procedure. The observed and calculated PDF's are compared with the help of the reliability indices. The refined parameters of LiNbO$_3$ using pair distribution function are given in table 5.3.1. Table 5.3.2 gives the nearest neighbour atom distances obtained from the PDF analysis for LiNbO$_3$. Figure 5.3.1 shows the fitted observed and calculated pair distribution function of LiNbO$_3$. In this figure, r(Å) represents the atomic distance from the origin and G(r) represents the reduced pair distribution function. Figure 5.3.1 shows the PDF peaks representing the interaction between Li–Li, Li–Nb, Li–O, Nb–Nb and O–O pairs. PDF peaks up to a distance of 20 Å are suitably indexed as shown in figure 5.3.1.

Table 5.3.1 The refined parameters from pair distribution function of LiNbO$_3$.

Parameter	LiNbO₃
Data range (Å)	0.0 – 25.0
Refinement range (Å)	3.0 – 20.0
Change in R	0.0018

R- Reliability index

Table 5.3.2 Nearest neighbour distances from PDF analysis of LiNbO₃.

Atom pair	Inter-atomic distances from PDF analysis (Å)		*Calculated Inter-atomic distances (Å)
	Observed	Calculated	
Li–Li	3.80	3.84	3.810
Li–Li	5.06	5.24	5.112
Li–Nb	6.02	-----	5.994
Li–Li	8.82	8.62	8.855
Li–Li	11.54	11.28	11.555
Nb–O	12.24	12.20	12.216
O–O	13.74	13.72	13.816
Li–O	15.44	15.40	15.423
Nb–O	16.88	16.90	16.721
Nb–O	19.54	19.58	19.691

*GRETEP for Windows (Jean Laugier and Bernard Bochu, 2002)

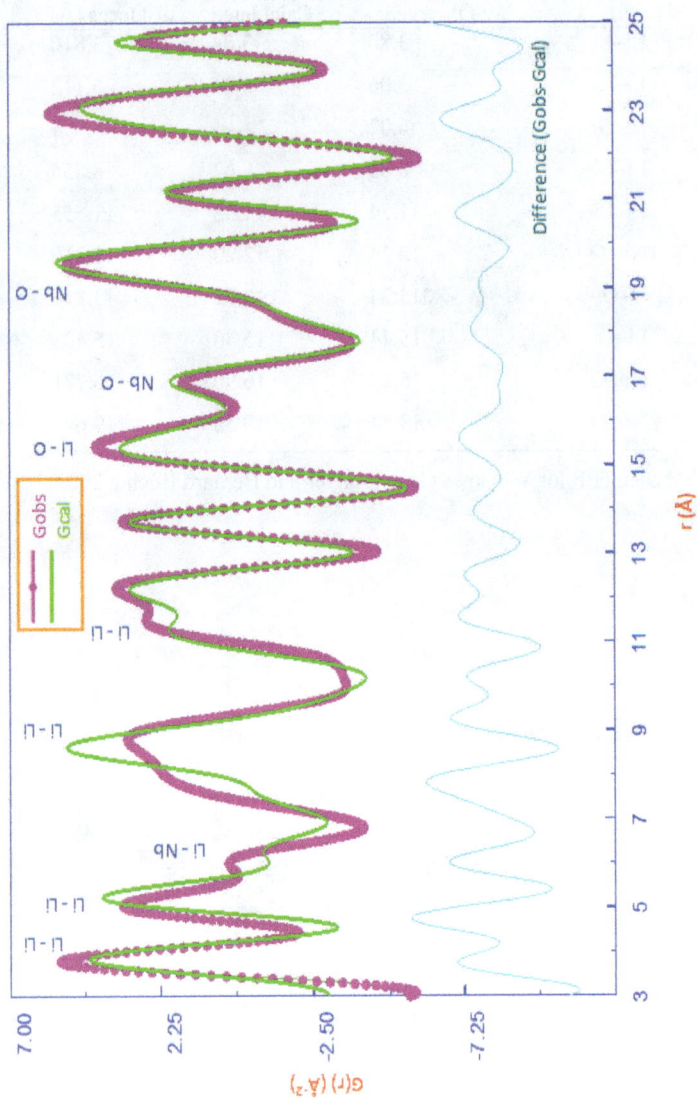

Figure 5.3.1 The fitted observed and calculated pair distribution function of LiNbO₃.

5.4 Conclusion

The observed PDF of $PbMoO_4$ and $LiNbO_3$ have been obtained from the raw intensities using a software program PDFgetX. A comparison between the observed and calculated PDF has been carried out using the software package PDFFIT. The refined pair distribution function for the samples $PbMoO_4$ and $LiNbO_3$ shows good matching of observed and calculated PDF's. The PDF refinement can be considered as equivalent to matching the observed X-ray powder data with too little structure parameters. Though the Q value of the powder X-ray data sets used in this doctoral work is small compared to the neutron or synchrotron data, an attempt has been successfully made for the study of atomic pair distribution of the non-linear optical materials $PbMoO_4$ and $LiNbO_3$.

References

[1]   Billinge S. J. L and Egami T., Phys. Rev. B, Vol. 47, pp. 14386, 1993.
https://doi.org/10.1103/PhysRevB.47.14386

[2]   Billinge S. J. L and Kanatzidis M. G., Chem. Commun., Vol. 7, pp. 749, 2004.
https://doi.org/10.1039/b309577k

[3]   deGroot F., Chemical Reviews, Vol. 101, pp. 1779, 2001.
https://doi.org/10.1021/cr9900681

[4]   Egami T., Materials transactions, JIM. Vol. 31(3), pp. 163, 1990

[5]   Farrow C. L., Juhas P., Liu J. W., Bryndin D., Bozin E. S., Bloch J., Proffen Th and Billinge S. J. L., J. Phys. Condens. Matter, Vol.19, pp. 335219, 2007.
https://doi.org/10.1088/0953-8984/19/33/335219

[6]   Filipponi A., Di Cicco A, C. R and Natoli C. R., Physical Review B, Vol.52(21), pp. 15122, 1995. https://doi.org/10.1103/PhysRevB.52.15122

[7]   Frey F., Acta Cryst. B, Vol. 51, pp. 592, 1995.
https://doi.org/10.1107/S0108768195002722

[8]   Jeong I. K., Thompson J., Proffen Th., Perez A and Billinge S. J. L., PDFGetX, A program for obtaining the Atomic Pair Distribution Function from X-ray Powder Diffraction Data, 2001

[9]   Jean Laugier and Bernard Bochu: GRETEP, Domaine universitaire BP 46, 38402 Saint Martin D'Heres http: /www.inpg.fr / LMGP/, 2002

[10]   Lytle F. W., Journal of Synchrotron Radiation, Vol. 6, pp. 123,1999.
https://doi.org/10.1107/S0909049599001260

[11] Neder R. B and Proffen T., Diffuse Scattering and Defect Structure Simulations : A Cook Book Using the Program DISCUS, Oxford university press, 2009

[12] Petkov V., Jeong I. K., Chung J. S., Thorpe M. F., Kycia S., Billinge S. J. L., Phys. Rev. Lett., Vol. 83, pp. 4089, 1999. https://doi.org/10.1103/PhysRevLett.83.4089

[13] Proffen T and Billinge S. J. L., J. Appl. Cryst., Vol.32, pp.572, 1999. https://doi.org/10.1107/S0021889899003532

[14] Rehr J. J and Albers R. C., Reviews of Modern Physics, Vol.72, pp. 621, 2001. https://doi.org/10.1103/RevModPhys.72.621

[15] Rietveld H. M., J. Appl. Cryst., Vol. 2, pp. 65, 1969. https://doi.org/10.1107/S0021889869006558

[16] Sayers D. E., Stern E. A and Lytle F.W., Physical Review Letters, Vol.27, pp. 1204, 1971. https://doi.org/10.1103/PhysRevLett.27.1204

[17] Stumm von Bordwehr R., Annales de Physique, Vol.14, pp. 377, 1999. https://doi.org/10.1051/anphys:01989001404037700

[18] Toby B. H and Egami T., Acta Cryst. A, Vol. 48, pp. 336, 1992. https://doi.org/10.1107/S0108767391011327

[19] Welberry T. R and Butler B. D., Chem. Rev. Vol. 95, pp. 2369, 1995. https://doi.org/10.1021/cr00039a005

Chapter 6

Conclusion

This research work concentrates towards the bonding behavior of the chosen non-linear optical materials through the electron density distribution studies. The electron density, bonding and charge transfer studies analysed in this work gives fruitful information to researchers in various fields. These properties can be utilized for the proper engineering of these technologically important non-linear optical materials.

1. Lead molybdate (PbMoO$_4$)

The structural parameters of PbMoO$_4$ have been refined from the powder X-ray diffraction data sets with the well-known Rietveld powder profile fitting methodology using the software package JANA 2006. The versatile tool MEM has been adapted to construct the electron density distribution in the unit cell. MEM charge density distribution has been visualized through the software VESTA. PbMoO$_4$ sample has been analyzed using SEM and UV-Visible spectroscopy for morphological and optical properties respectively. A comparison between the observed and calculated pair distribution function (PDF) has been carried out using the software package PDFfit. Pair distribution function (PDF) results give a clear picture of the local structure in PbMoO$_4$.

2. Lithium niobate (LiNbO$_3$)

Lithium niobate (LiNbO$_3$) crystal has been grown by Czochralski method. The refined structural information from powder X-ray diffraction (PXRD) data set of LiNbO$_3$ using Rietveld refinement technique has been used to construct the electron density distribution in the unit cell. The bonding feature of LiNbO$_3$ has been elucidated through 3D, 2D maps and 1D electron density profiles. The particle size, the elemental compositions and optical band gap analysis of LiNbO$_3$ has been carried out using SEM, EDS and UV-Visible spectroscopy respectively. The local structure of LiNbO$_3$ has been analysed for the nearest neighbor distances using pair distribution function (PDF).

3. Ce doped gadolinium gallium garnet (Ce:Gd$_3$Ga$_5$O$_{12}$)

Ce doped gadolinium gallium garnet (Ce:Gd$_3$Ga$_5$O$_{12}$) samples have been synthesized by sol-gel method. The structural parameters of Gd$_{3-x}$Ce$_x$Ga$_5$O$_{12}$ (x = 0.5, 1 and 3) have been refined using Rietveld refinement technique. The bonding behavior has been elucidated and analysed using MEM. The quantitative measurement of charge density and the variation of cell parameter with lanthanide doping reveals the inclusion of trivalent Ce^{3+}

dopant on the host lattice. Thus, the doping of Ce affects the charge ordering and hence the material's optical behavior in $Gd_3Ga_5O_{12}$. The particle size of the prepared $Ce:Gd_3Ga_5O_{12}$ samples have been estimated using SEM. The UV-Visible spectroscopy technique has been carried out to evaluate the band gap energy.

4.   Calcite (CaCO₃)

The electron density distribution in the unit cell of $CaCO_3$ has been constructed through MEM using the refined X-ray structure factors extracted from the Rietveld refinement technique. The charge density distribution in $CaCO_3$ shows the covalent along with ionic nature existing between Ca–O atoms and perfect covalent nature existing between C–O atoms. The morphological and optical band gap analysis has been carried using scanning electron microscopy and UV-Visible spectroscopy respectively.

5.   Yb doped calcium fluoride (Yb:CaF₂)

Yb doped calcium fluoride ($Ca_{1-x}Yb_xF_2$ (x=0.00, 0.03, 0.06, 0.09, 0.12)) materials have been synthesized through co-precipitation route. The powder X-ray diffraction (PXRD) data of undoped and Yb doped CaF_2 materials have been refined for observing the structural changes due to the inclusion of Yb using Rietveld refinement technique. No secondary phase has been observed. The refined X-ray structure factors have been used for the MEM procedure to obtain the charge density distribution in the unit cell. The electronic rearrangement and the bonding behavior in the unit cell of $Ca_{1-x}Yb_xF_2$ (x=0.00, 0.03, 0.06, 0.09, 0.12) has been analysed. At Yb 6%, the BCP (Bond Critical Point) decreases. An increase in particle size has also been observed from SEM micrographs of Yb doped calcium fluoride at Yb 6%.

6.   Al₂O₃, Cr:Al₂O₃ and V:Al₂O₃

The undoped and doped Al_2O_3 materials have been characterized by PXRD for analyzing structural information using Rietveld refinement technique. The electron density distribution and the observed charge ordering behavior with the inclusion of Cr at 5% and V at 5% in the unit cell of Al_2O_3 have been analysed through MEM method. Further analyses have been done through SEM, optical and other compositional studies. The preliminary results indicate that it is possible to distinguish the doping effects in Al_2O_3 through the electron density analysis. The refined powder profiles, the observed and calculated structure factors, the 3D electron density iso-surfaces, the 2D electron density contour maps, the 1D electron density profiles and the SEM pictures- all support the addition of Cr and V atoms in the host matrix of Al_2O_3.

Keyword Index

About the Author

Dr Ramachandran Saravanan, has been associated with the Department of Physics, The Madura College, affiliated with the Madurai Kamaraj University, Madurai, Tamil Nadu, India from the year 2000. He is the head of the Research Centre and PG department of Physics. He worked as a research associate during 1998 at the Institute of Materials Research, Tohoku University, Sendai, Japan and then as a visiting researcher at Centre for Interdisciplinary Research, Tohoku University, Sendai, Japan up to 2000.

Earlier, he was awarded the Senior Research Fellowship by CSIR, New Delhi, India, during Mar. 1991 - Feb.1993; awarded Research Associateship by CSIR, New Delhi, during 1994 – 1997. Then, he was awarded a Research Associateship again by CSIR, New Delhi, during 1997- 1998. Later he was awarded the Matsumae International Foundation Fellowship in1998 (Japan) for doing research at a Japanese Research Institute (not availed by him due to the simultaneous occurrence of other Japanese employment).

He has guided eleven Ph.D. scholars as of 2017, and about five researchers are working under his guidance on various research topics in materials science, crystallography and condensed matter physics. He has published around 140 research articles in reputed Journals, mostly International, apart from around 50 presentations in conferences, seminars and symposia. He has also guided around 60 M.Phil. scholars and an equal number of PG students for their projects. He has attracted government funding in India, in the form of Research Projects. He has completed two CSIR (Council of Scientific and Industrial Research, Govt. of India), one UGC (University Grants Commission, India) and one DRDO (Defense Research and Development Organization, India) research projects successfully and is proposing various projects to Government funding agencies like CSIR, UGC and DST.

He has written 8 books in the form of research monographs including; "Experimental Charge Density - Semiconductors, oxides and fluorides" (ISBN-13: 978-3-8383-8816-8; ISBN-10:3-8383-8816-X), "Experimental Charge Density - Dilute Magnetic Semiconducting (DMS) materials" (ISBN-13: 978-3-8383-9666-8; ISBN-10: 3-8383-9666-9) and "Metal and Alloy Bonding - An Experimental Analysis" (ISBN -13: 978-1-4471-2203-6). He has committed to write several books in the near future.

His expertise includes various experimental activities in crystal growth, materials science, crystallographic, condensed matter physics techniques and tools as in slow evaporation, gel, high temperature melt growth, Bridgman methods, CZ Growth, high vacuum sealing etc. He and his group are familiar with various equipment such as: different types of cameras; Laue, oscillation, powder, precession cameras; Manual 4-circle X-ray diffractometer, Rigaku 4-circle automatic single crystal diffractometer, AFC-5R and AFC-7R automatic single crystal diffractometers, CAD-4 automatic single crystal diffractometer, crystal pulling instruments, and other crystallographic, material science related instruments. He and his group have sound computational capabilities on different types of computers such as: IBM – PC, Cyber180/830A – Mainframe, SX-4 Supercomputing system – Mainframe. He is familiar with various kind of software related to crystallography and materials science. He has written many computer software programs himself as well. Around twenty of his programs (both DOS and GUI versions) have been included in the SINCRIS software database of the International Union of Crystallography.